Buddhism and Bioethics

Also by Damien Keown and from the same publishers

THE NATURE OF BUDDHIST ETHICS

Buddhism and Bioethics

Damien Keown
Reader in Buddhism
Goldsmiths College
University of London

No paragraph of this publication may be reproduced, copied or
transmitted save with written permission or in accordance with
the provisions of the Copyright, Designs and Patents Act 1988,
or under the terms of any licence permitting limited copying
issued by the Copyright Licensing Agency, 90 Tottenham Court
Road, London W1P 0LP.

Any person who does any unauthorised act in relation to this
publication may be liable to criminal prosecution and civil
claims for damages.

The author has asserted his right to be identified
as the author of this work in accordance with the
Copyright, Designs and Patents Act 1988.

First edition 1995
Reprinted (with new Preface and minor alterations) 2001

Published in paperback 2001 by
PALGRAVE
Houndmills, Basingstoke, Hampshire RG21 6XS and
175 Fifth Avenue, New York, N.Y. 10010
Companies and representatives throughout the world

PALGRAVE is the new global academic imprint of
St. Martin's Press LLC Scholarly and Reference Division and
Palgrave Publishers Ltd (formerly Macmillan Press Ltd).

ISBN 0-333-61858-0 hardback (outside North America)
ISBN 0-312-12671-9 hardback (in North America)
ISBN 0-333-91280-2 paperback (worldwide)

This book is printed on paper suitable for recycling and
made from fully managed and sustained forest sources.

A catalogue record for this book is available
from the British Library.

The Library of Congress has cataloged the hardcover edition as follows:
Keown, Damien, 1951–
    Buddhism & bioethics / Damien Keown.
       xv, 208 p. ; 23 cm.
    Includes bibliographical references (p. 199-203) and indexes.
    ISBN 0-312-12671-9
       1. Buddhist ethics. 2. Bioethics. I. Title: Buddhism and bioethics.
    II. Title.
    BJ1289 .K43 1995
    294.3'5642—dc20
                                                            95-005776

Transferred to digital printing 2006

For Jo

One should respect the supreme value and sacredness of life

Asoka (Brahmagiri Rock Edict II) *c.*250 BC

Contents

Preface to the
Paperback Edition

Since the first hardback edition of this book was published in 1995 interest in the subjects discussed in these pages has become even more intense as the pace of medical and scientific developments has accelerated and as public awareness of the new technologies has continued to grow. Abortion and euthanasia remain, as ever, highly controversial issues but have been supplemented by a range of new problems connected with genetic engineering. Were I writing the book afresh I would include a discussion of the issues raised by cloning and related matters, but it is too late to include them now and I propose to reserve them for separate discussion in a future publication. In other respects, however, the topics covered remain topical and relevant both to Buddhists and outsiders at the threshold of the new millennium.

It is also pleasing to see an increasing number of researchers turning their attention to problems and issues in Buddhist ethics. Along with recent books on the subject there has existed, since 1994, a new forum in the online *Journal of Buddhist Ethics* (http://jbe.la.psu.edu) which I co-edit with my colleague Charles Prebish. The electronic format of this publication on the world wide web serves to promote research in this area and to allow the results to be disseminated to the academic community and beyond.

This paperback edition has given me the opportunity to correct minor typographical errors, but is in all other respects unchanged from the original hardback edition.

Introduction

This book is an attempt to apply Buddhist principles to some major contemporary problems in biomedical ethics. It is the first contribution of its kind and is written for a broad general readership ranging from specialists in Buddhism who may know little about medical ethics to ethicists with an interest in medical issues who know little or nothing of Buddhism. It will also be of interest to the growing number of Buddhists in the West and elsewhere who would like to see these issues receive a higher priority than they have been given so far either by the tradition or the academic community.

My intention has been to make the book accessible to the widest possible audience by expressing Buddhist ideas and concepts in a form intelligible to the general reader. It assumes no prior knowledge of either Buddhism or ethics. Technical terms and foreign words have been kept to a minimum, and English equivalents used wherever possible. Diacritical marks have been omitted with the exception of the tilde (~) which functions as an aid to pronunciation. A brief sketch of Buddhism is given below and a glossary of Buddhist terms is provided at the end.

Despite the contemporary importance of issues such as abortion and euthanasia, there has been comparatively little discussion of them from a Buddhist perspective. Other problems in medical ethics such as embryo research and the definition of 'death' have scarcely been raised. My intention is to explore this archipelago of little-known moral islands and in so doing to construct an intellectual bridge of some kind between them. In more prosaic terms the aim is to formulate a set of principles which can be applied with consistency across a range of biomedical problems. In terms of the academic discipline involved we understand the present study as a contribution to the field of *applied normative ethics*.[1] As the twenty-first century fills the horizon, one of the greatest challenges facing Buddhism is to evolve a perspective on bioethical issues which is both comprehensive and systematic. It is beyond the scope of any single study to achieve both of these goals, and the present volume may be thought of as a prologue to the work which lies ahead. Each of the substantive issues dealt with here deserves at least a volume to itself, and it is to be hoped that the opportunity will arise in due course to provide a more selective treatment of the myriad issues here left unresolved.

Buddhism: a thumbnail sketch

What is known to the West as 'Buddhism' is a body of religious teachings attributed to an historical individual who lived in North East India in the fifth century BC. Following a profound spiritual transformation achieved at the age of thirty-five he became known by the honorific title of 'Buddha' ('enlightened one'). The Buddha claimed no divine provenance for his teachings and understood them as being grounded in the nature of things. For this reason the word for the teachings (Dharma) also refers to the immutable laws of both the natural and moral orders of which they are the expression. Personal realisation in Buddhism, therefore, consists of living in accordance with Dharma, and anyone who follows Buddhist teachings can replicate the spiritual transformation achieved by the founder. The teachings are expounded in the form of four basic propositions known as the Four Noble Truths. These maintain that life as we now know it is imperfect and unsatisfactory; that the causes of this unsatisfactoriness are craving and ignorance; that there exists a state of perfection free from all deficiencies (nirvana); and that the way to perfection is the Eightfold Path. The Eightfold Path is a programme for right living which emphasises three things: moral cultivation, meditation, and knowledge of the true nature of the human condition. As far as moral conduct is concerned the Buddha laid down certain basic precepts. There are many formulations of precepts in Buddhism but the best known are the Five Precepts for laymen. The Five Precepts forbid:

1. Taking life
2. Stealing
3. Sexual misconduct
4. Lying
5. Taking intoxicants

The Buddha appointed no successor, and many different schools arose after his death. Buddhism spread widely and influenced every Asian civilization, but it has no head and there is no central authority which is the custodian of orthodoxy. The Buddhism of southern Asia is predominantly that of the more conservative Theravada school, while that of northern Asia belongs to the more doctrinally innovative movement known as Mahayana. The views expressed in this book are based on the canonical and commentarial literature of the Theravada school. These sources, recorded in the Pali language, are the closest we are likely to get to the ethical teachings of the Buddha, and I will use them as a touchstone for validating views, opinions and arguments. Whether or not these sources reliably record the Buddha's teachings they are among the most ancient and are at least as authoritative as any other.

The canonical scripture of the Theravada school is contained in a collection known as the Pali Canon. This consists of three separate collections of texts: the Discourses (*sutta*), which for the most part are teachings and sermons given by the Buddha; the Monastic Rule (*vinaya*), which contains the ethico-legal rules which regulate the conduct of the Order (*sangha*) and its members; and the Scholastic Treatises (*abhidhamma*), which are slightly later texts devoted to the analysis and classification of the teachings. According to tradition the canon was fixed at the First Council, which took place shortly after the Buddha's death. The texts themselves were not committed to writing until the first century BC. The commentarial literature on the canon is extensive, but the most influential commentaries are those attributed to a monk named Buddhaghosa, who flourished in the fifth century AD in Sri Lanka. In terms of status he may be thought of as the Buddhist Aquinas. For our present purposes the sources which are of most relevance are the Monastic Rule and Buddhaghosa's commentary upon it. The textual sources used are the various editions of the Pali Text Society, and abbreviations follow the style of *The Pali Text Society's Pali–English Dictionary*. Translations are my own unless stated otherwise.

Aristotle and natural law

We do not begin our enquiry entirely without direction, and the approach adopted here is based on the conclusions concerning the theoretical basis of Buddhist ethics reached in *The Nature of Buddhist Ethics*. It was suggested there that Buddhism is best understood as a teleological virtue ethic. This means that Buddhism postulates a certain goal or end (*telos*) as the fulfilment of human potential, and maintains that this goal is to be realised through the cultivation of particular practices. In Buddhist terms, the goal of human perfection is nirvana and this is attained through the process of moral and intellectual self-transformation which comes about through following the Eightfold Path.

The present work takes this conclusion as its starting point and asks how these theoretical principles would be applied in practice to specific biomedical problems. The intellectual framework within which the issues are addressed may therefore be described as Aristotelian. The ethical principles of Aristotle form the cornerstone of a tradition of moral reflection which has developed under the name of 'natural law'. Although Aristotle looms large in this tradition it may be thought of as a rope made up of many strands, with contributions from Greek, Roman and Christian thought. Natural law reflection begins with the question: 'What is it for a human being to flourish?' The transcultural phenomenon we refer to as 'Buddhism' is concerned essentially with the same question, and the natural law tradition provides an illuminating Western parallel which can be helpful in understanding Buddhism. The leading contemporary exponents of natural law, philosophers such as Germain Grisez and John Finnis, have shown how its principles can be applied to moral issues in many areas, including those in the field of medicine. Their approach provides, in certain respects, an interesting analogue to Buddhism, and suggests fruitful points of contact. Despite the similarities, however, there remain many aspects of Buddhist ethics which are problematic. The Buddhist moral perspective differs from the Western in many respects, not least in its belief in reincarnation or rebirth,[2] and its less pronounced distinction between human and non-human species. These differences create significant complications and give theoretical reflection on Buddhist ethics something of the flavour of three-dimensional chess.

East and West

There are many contrasts inherent in the application of Buddhist principles to biomedical ethics. Three in particular may be singled out. The first can be described as chronological, in that we are seeking to apply some of the world's earliest beliefs to the latest ethical problems. The second is cultural, in that we are interrogating an Eastern culture for answers to problems that have arisen in the West, due not least to technological developments. The third, which is not unrelated to the second, may be described as ideological, in that we are endeavouring to apply religious beliefs and values in contexts where the relevant vocabulary is predominantly that of materialist science. There is no accepted methodology for dealing with such a complex set of contrasts. Indeed, the study of Buddhist ethics itself has hardly begun and there are fundamental issues in the field of comparative ethics yet to be resolved. Add to this the inherent complexities in the subject matter of medical ethics and the difficulty in finding consensus at any level, and one could be forgiven for feeling that there is little hope that a credible 'Buddhist' position can be articulated at this stage.

In one respect such pessimism may be exaggerated. Buddhism itself would not wish to make a radical distinction between 'ancient' and 'modern' problems. It believes that all dilemmas – old and new – can be successfully analysed and resolved by reference to the eternal moral law (Dharma). Moreover, in contrast to the endemic relativism of much modern thought, Buddhism would not accept that the principles of Dharma vary from one culture to another. It would also question the necessity for any ideological clash between religion and science, arguing that both are intellectual structures which aim at the discovery of truth about the nature of man and the universe he inhabits. On this view there is every reason to expect them to converge rather than remain separate. For these reasons Buddhism may feel itself free and qualified without further preliminaries to confront the ethical challenges which arise from its encounter with the West. If one sets out simply to act as an interpreter of Buddhism and the principles which inform it, then, perhaps there are no insuperable methodological obstacles in the way of applying Buddhist principles directly to modern dilemmas.

What is 'Buddhism'?

Looking at the tradition from the outside, however, certain problems remain. Not the least of these is the question: what is 'Buddhism'? When we seek to apply the ethical principles of Buddhism, precisely which *form* of Buddhism do we have in mind? Is it the Buddhism of India, or Tibet, or China, or Japan, or South-East Asia? Furthermore, within any one of these cultural traditions there may be a variety of sects and schools, so which one of these is to be chosen as representing the 'Buddhist' view? If there is to be a discipline of Buddhist ethics it seems imperative there be some ground rules for determining what can count as a 'Buddhist' position. A further complication is that many Westerners who have adopted Buddhism have begun to formulate and express their own views as to how Buddhism should respond to the challenges of contemporary life. What is the status of these views? Are they 'Buddhist' or some new hybrid which lacks an authentic pedigree? As we approach the twenty-first century an increasingly urgent question will be: 'Whose is the authentic voice of Buddhism?'

Methodological issues

Although Buddhism may feel equal to the challenge of contemporary medical ethics, the procedures that should govern the academic study of Buddhism itself are another matter. A range of problems arise here: for example, on what evidence should conclusions about Buddhist ethics be based? Is scripture the only source of valid knowledge, or must attention also be paid to anthropological evidence concerning beliefs and practices in Buddhist societies? An even larger question is whether either of these forms of evidence requires decoding or 'deconstructing' against the cultural background in which they arise. For example, it may be thought that certain 'doctrines' (e.g. the belief in rebirth) found their way into Buddhism only because they were part of a particular cultural milieu. Since a belief in rebirth is not part of traditional Western culture, can it now be reinterpreted or perhaps jettisoned altogether? Textual sources, finally, are by nature always historically and culturally conditioned, yet throughout this book I will appeal to them as authoritative evidence for the 'Buddhist view' on ethical issues and seek to apply them in a modern context. Is this legitimate, or are these sources too 'culture-bound' to be of any normative value today?

Buddhist fundamentalism

This reliance on the authority of ancient texts gives the book a fundamentalist flavour, an impression which will be reinforced by the conservative nature of the conclusions reached. By 'fundamentalism' here is not meant emotional, anti-intellectual fanaticism, but the requirement that views and opinions be grounded in textual sources.[3] If the essence of fundamentalism is deference to scripture, then such a characterisation may not be inappropriate with certain qualifications. The first is that fundamentalism, in a Buddhist context, does not entail belief that every line of scripture is incontrovertible truth; that all truth is contained in scripture, or that scripture is the only source of truth. Buddhist fundamentalism asserts only that moral truth *can be found* in scripture. A further qualification is that fundamentalism does not preclude the adoption of a text-critical methodology. To recover the meaning of a text is not always an easy task, and the sense can be corrupted in many ways through scribal error, the interpolation of later material, and the arbitrary shuffling of passages. Nor can commentaries always be relied on since they are sometimes given to fanciful interpretations when the original meaning has clearly been lost. What we understand as 'Buddhist fundamentalism', then, if this label has any use as a methodological designation, is an approach to the study of Buddhism which holds that if the problems of scriptural interpretation or 'hermeneutics' can be overcome, and the meaning of scripture established beyond reasonable doubt, then what has been recovered are moral truths which are as valid today as they were in the fourth century BC. Other philosophical teachings found in Buddhist sources, such as those concerning personal identity and causation, are universally regarded in this way, and we see no reason why its moral teachings should not be also.

Six questions

Thirty years ago in a book well ahead of its time, Winston King raised six important questions for Buddhist ethics.[4] Few answers have been forthcoming.[5] In connection with our present subject six alternative questions arise, most of which have been raised but not answered in the discussion so far. Is a special methodology required for cross-cultural ethics, and if so, what might it be? Is there a 'Buddhist view' on ethical issues, or only the views of individual texts, schools, teachers and practitioners? What role should scripture play in establishing moral norms? Does Buddhism have fundamental moral principles (moral absolutes) or does it hold that what is right varies according to 'the situation'? What is the moral status of animals and other forms of non-human life? What is the proper role of compassion in the moral life? Questions of this kind will provide the agenda for Buddhist ethics in the years ahead. No pretence is made that they will receive a thorough or systematic airing here, although we will respond in a preliminary way to some of them in Chapter 1. The reflections there will guide our thinking in relation to the substantive ethical issues discussed in the remainder of the book. Chapter 2 is devoted to moral issues which cluster around the beginning of life, and Chapter 3 deals with moral issues surrounding death and dying.

Acknowledgements

I wish to record my thanks to my brother John, Lecturer in the Law and Ethics of Medicine at Cambridge University, for generously sharing his expertise in discussions too numerous to recall, and for his comments and suggestions on drafts of the manuscript. My thanks also to Jason Lyon for his diligent proofreading. I remain, of course, solely responsible for the accuracy of what follows.

1

Buddhism, Medicine and Ethics

Introduction

Since the subject of this book is ethics in the context of medicine, it will be appropriate to begin with a few reflections on the links between Buddhism and medical practice. We will then turn in Section II to the role of ethics in Buddhism as an historical tradition. This will lead on in Section III to a consideration of how the subject of Buddhist ethics is to be approached in the present work. Section IV examines issues concerning moral subjects and Section V contains a discussion of the relation between ethics and human good in Buddhism. Section VI concludes the chapter by applying the theoretical conclusions reached to a case-history from the Monastic Rule.

I BUDDHISM AND MEDICINE

R.L. Soni has written, 'It is indeed a matter of supreme interest that the noble profession of medicine and the corpus of thought known as Buddhism are both concerned in their own way in the alleviation, control and ultimately the removal of human sufferings.'[1] In a similar vein, under its entry on 'Buddhism' the *Dictionary of Medical Ethics* points out that 'The principles governing Buddhism and the practice of medicine have much in common.'[2] If we seek a doctrinal basis for the link between medical practice and Buddhist doctrine we will find it in the Four Noble Truths. It is under the First Noble Truth that the Buddha sets out the basic problem faced by mankind.

The First Noble Truth

The First Noble Truth points out that all forms of embodied existence are unsatisfactory by virtue of the physical and mental

1

suffering which is inherent in them. It states: 'Birth is suffering, sickness is suffering, old age is suffering, death is suffering; pain, grief, sorrow, despair and lamentation are suffering.' The four physical aspects of suffering mentioned, namely birth, sickness, old age and death, may involve physical pain to a greater or lesser degree. When Buddhism characterises these experiences as 'suffering', however, it means more than that they are 'painful'. The word translated as 'suffering' (*dukkha*) includes physical pain, but denotes more broadly the profound unsatisfactoriness of the very mode of being within which birth and death occur. This unsatisfactoriness stems from the fact that existence as we know it is constantly exposed to the possibility or risk of pain in the situations described. Seen against the background of the doctrines of karma and rebirth, it is the unavoidability of *repeated* birth and death with its attendant physical discomfort to which the First Noble Truth draws attention. In the long cycle of lifetimes which, according to Buddhism, all experience, no one can expect that their lives will remain free of pain and disease. Whatever advances are made by medical science it is unlikely that there will be a cure for every complaint. No one is immune from illness, and even the Buddha received medical treatment during his lifetime. Apart from disease there is always the risk of accidents. In the final analysis it is unlikely that medical science will ever conquer sickness or death, though it may succeed in extending the average lifespan far beyond its present limits. The *psychological* problems mentioned under the First Noble Truth are perhaps even more intractable, and conditions involving anxiety and depression can be more debilitating and difficult to treat than physical infirmities. The point need not be laboured, and the extensive catalogue of human mental and physical afflictions is well known to physicians and laymen alike.

Buddhism draws attention to the shortcomings of human existence not out of a morbid fascination with suffering but in order to encourage a realistic appraisal of the human condition. It is not until the condition has been accurately diagnosed that the search for a remedy can begin. Suffering is not something to be relished in a perverse or masochistic sort of way – quite the contrary. Good health and freedom from pain are important aspects of human well-being and are highly valued by Buddhism. Its ultimate goal, however, is a *permanent* cure for life's afflictions, something which cannot be achieved through medicine alone.

Medicine in India

The Buddhist monastic Order (*sangha*) has a claim to be the world's oldest and most widespread continuous social institution. For over two thousand years it has, amongst its other activities, maintained a close involvement with the treatment of the sick. Several centuries before Christ, Buddhist monks were developing treatments for many kinds of medical conditions, and, according to the latest research, Buddhism can claim much of the credit for the development of traditional Indian medicine (*Ayurveda*).[3]

Kenneth Zysk has recently described the paradigm shift which occurred in Indian medicine in the period between 800 and 100 BC involving a change from the older magico-religious healing techniques to a 'an entirely new empirico-rational approach to disease and its cure'.[4] The Buddha raised a dissenting voice to the contemporary religious orthodoxy which went by the name of 'Brahmanism', and his followers were not hindered by the restrictions on medical research which arose from orthodox beliefs concerning the impurity of death and disease. To the orthodox Brahmin, bodily fluids such as blood represented a potent source of ritual pollution, and the status of the physician was accordingly low. Indeed, the first dissection by an Indian medical student was not performed until 1836, when the government fired a salute of guns from Fort William in Calcutta to honour the occasion. In ancient times those who sought empirical knowledge of the body and its functions were socially marginalised, and found themselves in the company of other unorthodox groups such as Buddhist monks.

Medicine and monasticism

According to Zysk, the early Buddhist monasteries of India were the places where the most significant developments in Indian medicine took place.

> Like the Christian monasteries and nunneries of the European Middle Ages, communities of Buddhist monks and nuns played a significant role in the institutionalization of medicine ... The codification of medical practices within the monastic rules accomplished perhaps the first systematization of Indian medical knowledge and probably provided the model for later handbooks of medical practice; the monk-healers' extension of

medical care to the populace and the appearance of specialized monastic structures serving as hospices and infirmaries ... ensured ongoing support of the monasteries by the laity; and the integration of medicine into the curricula of major monastic universities made it a scholastic discipline. In India and elsewhere in Asia, Buddhism throughout its history maintained a close relationship with the healing arts, held healers in high esteem, and perhaps best exemplified the efficacious blending of medicine and religion.[5]

In spite of this close relationship there were, particularly in the early period, restrictions designed to deter monks from taking a professional interest in medicine.[6] Specific medical practices are singled out in the early sources as inappropriate ways for monks to earn a livelihood. Those mentioned include the administering of purgatives and emetics, treatment of the ears, eyes and nose, and surgery and paediatrics.[7] Birnbaum correctly interprets these restrictions as 'a warning against habitual treatment of laymen (especially for the sake of alms), a warning against becoming a doctor rather than devoting time to the spiritual exercises of early Buddhist practices.'[8] Medicine was considered a secular art, and monks had a prior commitment to a vocation with other priorities. At the same time, medical expertise was required as a means to securing the healthy physical constitution necessary to withstand the rigours of the monastic life. Without good health, as Birnbaum points out, the practice of the religious life would have been impossible.

The four requisites for life, stated repeatedly in the various texts of the Pali Canon, are robes, food, lodging, and medicine. It is not surprising that medicine bears such significance, for surely great strains were placed upon the physical well-being of monks due to their austere life and strenuous meditative practices. Since illness and its indisposition tend to weaken the mind, often causing it to lose its focus on its function as a liberating faculty, the prevention and proper treatment of illness held (and continues to hold) a great importance for the Buddhist monk.[9]

The first beneficiaries of Buddhist medical expertise were therefore monks themselves. The Buddha pointed out that since monks had severed all other social ties it was incumbent on them to care for one another:

You, O monks, have neither a father nor a mother who could nurse you. If, O monks, you do not nurse one another, who, then, will nurse you? Whoever, O monks, would nurse me, he should nurse the sick.[10]

With the passage of time restrictions on treating the laity were eased. The great Buddhist monarch Asoka claims, in an edict around 258 BC, to have instituted an early form of state health-care provision:

Everywhere in the dominions of King Priyadarsi (i.e. Asoka) ... provision has been made for two kinds of medical treatment, treatment for men and for animals. Medicinal herbs, suitable for men and animals, have been imported and planted wherever they were not previously available. Also, where roots and fruits were lacking, they have been imported and planted.[11]

The inscriptions, unfortunately, do not record who delivered the medical care in question. Asoka's interest in medicine may have been stimulated by his conversion to Buddhism, and Buddhist monks may well have had some role to play in his 'national health service' if indeed it involved anything more than the planting of herbs and the like. What is certain, however, is that his royal endorsement of medical provision would have provided a further stimulus to medicine in the monasteries. As Buddhism spread, moreover, the goodwill generated by the provision of medical care would doubtless have encouraged monks to develop their skills in this area. Birnbaum sums up neatly the three reasons why Buddhist monks might take an interest in medicine: 'Thus a monk might learn healing techniques to aid his fellow monks, to be of compassionate service to laymen, and as an expedient means for obtaining trust for the purpose of spreading the Buddhist teachings.'[12] These reasons still apply, and it is not uncommon in the present day to find monks qualified in traditional medicine, Western medicine or both.

Given the close connection between medicine and monasticism, it will come as no surprise to find that the Buddhist attitude to the treatment and care of patients is deeply influenced by its religious beliefs. What is to be done and not to be done by the physician will be determined by the same moral principles which determine what is and what is not to be done by a monk, since the physician is a monk first and a physician second. Thus, as we might expect, medical ethics in Buddhism involves essentially the application of the wider principles of religious ethics to problems in a more specialised field.

The ancient monastic texts reveal that the Buddha resolved problematic matters of monastic discipline on a case-by-case basis as new situations arose. As such he was often asked to rule on the legitimacy of certain forms of treatment where these seemed to involve the infringement of a monastic rule. For example, there is a rule that monks should not eat after midday; but what if a monk became sick and could take no food in the morning? Rules of this kind were seen to stand in need of modification in the light of circumstances, and many exceptions were allowed where treatment of the sick was concerned. The record of these case-histories represents the earliest codification of medical knowledge in India,[13] and we will have cause to make reference to some of them in detail at various points in our discussion. As may be expected, the treatments described are not those of modern medicine, nor are the problems they raise identical in all respects. Given the primitive technology, for example, certain questions which have arisen today could scarcely have been imagined at the time. Nevertheless, we are not entirely bereft of guidance in the ancient sources; although the circumstances today may be new, the moral issues which arise turn out on analysis to be fundamentally the same. Our task, therefore, is to make explicit the principles which underlie the judgements in the ancient texts and apply them faithfully in a modern context. The problems involved in doing this will occupy our attention for the rest of the chapter.

II   BUDDHISM AND ETHICS

Buddhism is widely respected for its benevolent and humane moral values. What is lacking in the primary and commentarial sources, however, is a systematic exposition of the theoretical framework in terms of which Buddhist ethics is to be understood. This means that we are at something of a loss when we seek solutions to new problems for which there is no scriptural paradigm, or when two or more values seem to be in conflict. This problem will come increasingly to the fore as the encounter between Buddhism and the modern West gathers pace.

Buddhism and the West

Many Westerners who turn to Buddhism find that their experience falls into two stages. In the first stage they find that Buddhism speaks to their spiritual needs in a direct, refreshing and practical way, in contrast to what many regard as the excessively rigid and authoritarian attitude of religion in the West. The second stage begins later when they attempt to apply the teachings they have learned to practical problems they encounter in their lives. Here they find Buddhism less helpful, and unable to respond readily to the searching ethical questions they direct to it. In part, this is because many of the problems encountered today arise in a context which is unfamiliar to traditional Buddhism. It should be remembered that one of the major centres of Buddhist learning, Tibet, was to all intents and purposes a medieval feudal state as recently as half a century ago. Elsewhere in Buddhist Asia there was nothing to compare with the revolutionary developments such as the Enlightenment and the rise of science which swept away medievalism in Europe. To put it bluntly, Buddhism is a third-world phenomenon and several hundred years out of date. Even in those Asian countries which have seen rapid modernisation, such as Japan, it remains to be seen whether the arranged marriage between East and West will be a fruitful union. There must be some doubt, then, as to how successfully Buddhism can respond to the urgent demands of Westerners for guidance on issues about which it has little practical experience.

Perhaps this explains why the voice of Buddhism is so little heard in contemporary moral debate. Churchmen of all denominations contribute to discussions on every issue from Aids to Zygotes, and it is only natural that Western Buddhists should enquire as to the Buddhist position on these issues. To take a few examples from the field of bioethics: at what point does life begin and end? Is Buddhism in favour of experimentation on embryos and up to what stage? Does Buddhism permit abortion, and under what circumstances? Is euthanasia ever justified? Should a life-support machine ever be switched off, and if so, when? A still broader group of questions relate to the kind of world Buddhism would like to see, and its attitudes to society and the environment. Some of these issues have begun to be addressed in recent years from within a loose movement which describes itself as 'socially engaged Buddhism'. This movement consists mainly, but not exclusively, of Westerners who have sought to apply Buddhist principles to the problems of life in the modern world. Some progress is therefore being made, but it is difficult to disagree with Kosho Mizutani when he writes: 'I submit that a study of Buddhism that emphasises its ethical aspects will be the most important task facing Buddhists in the twenty-first r .tury.'[14]

The silence of the tradition

Very few of the issues listed above have received attention from Buddhist teachers or scholars. Why is this? One of the main reasons would seem to be the lack of precedent set by the tradition. It must be remembered that Buddhism originated as a movement whose purpose was to *renounce* social life, not to become enmeshed in its problems. The household life is depicted in the early sources as full of cares and burdens and contrasted with the freedom of the monk who has renounced worldly things.[15] The early members of the Buddhist Order were what Dumont has termed 'outworldly' individuals whose concern was with spiritual development rather than social reform. He writes:

> What is essential for us is the yawning gap between the renouncer on the one hand and the social world and the individual-in-the-world on the other. To begin with, the path of liberation is open only to those who leave the world. Distance from the social world is the condition for individual spiritual development. *Relativisation* of life in the world results immediately from world renunciation. Only Westerners could mistakenly suppose that some sects of renouncers would have tried to change the social order.[16]

Yet in spite of this initial impetus towards world-renunciation, Buddhism quickly came to envisage the relationship between monks and laymen as symbiotic: in return for material sustenance, monks should provide religious teachings. Surely, it might be thought, within the context of religious teachings should come moral guidance? Apart from the transmission of the precepts, however, little effort was invested in exploring the presuppositions of Buddhist ethics. By and large, the Buddha counselled laymen to follow the precepts he laid down, and he avoided discussion of theoretical problems. With few exceptions, this has remained the pattern.

Ethics in India

A second reason why Buddhism has had comparatively little to say on ethics may be that Indian culture as a whole has shown little interest in the subject as an independent philosophical discipline. Hindu ethics is concerned primarily with defining the duties of the different castes and their mutual relations. The concept of ethics as concerned with individual responsibility and personal choice is lost sight of against the background of the hierarchical structure of the caste system. An important principle of ethics in the West has been that the same moral rules should apply to one and all. In terms of the caste system such an idea is nonsense, for it is caste which determines the rightness of actions: thus the same act may be right for a Brahmin but wrong for an outcaste. Where Western ethics is founded on universalisation Hindu ethics is founded on particularisation.

The Buddha rejected the whole notion of caste with its segregation by birth and occupation, and its obsession with the classification of persons, things and acts as pure or impure according to the demands of ritual law. In the light of this, his reluctance to become involved in stipulating detailed rules of social conduct is understandable. His purpose in rejecting the caste system was not to erect an alternative social structure in its place, but to leave himself and his followers free to pursue spiritual goals. Accordingly, there never developed in Buddhism a science of religious law of the kind found in Hinduism, Judaism, Islam and Christianity. In each of these traditions jurists and commentators have established codes and digests of laws in a systematic attempt to resolve conflicts between daily life and the demands of sacred law. Since Buddhism was not concerned with the detailed regulation of lay society, however, there was no stimulus to the type of ethico-legal reflection from which moral philosophy is born.

As the Order flourished as an institution, however, questions concerning the regulation of communal life and its relations with society at large came increasingly to the fore. In the part of the Buddhist canon known as the Monastic Rule (*vinaya*) we find recorded some of the moral dilemmas thrown up by these developments. The Monastic Rule is a vitally important source of information since it is the only part of the canon where the Buddha is shown giving systematic judgements on specific cases as opposed to general moral teachings at the preceptual level. We shall have cause to make reference to it on several occasions.

Buddhism outside India

As Buddhism moved beyond India it found little incentive to develop a social philosophy of its own. When Buddhism reached China in the first century of the Christian era, it encountered a strong social order already in place. Confucianism was authoritative in matters of social conduct and Buddhism could not compete as a rival in this field. Instead it found a niche for itself as the third of the 'three religions' and offered its expertise in those areas where Confucianism and Taoism were weakest. In Tibet, on the other hand, where Buddhism encountered an unsophisticated feudal society, it quickly established itself as the dominant ideology and had no need to expand its traditional philosophical base. While recognising that we are speaking in the most general terms here, it might be said that because of a series of historical accidents Buddhism has never needed to address the kinds of issues which we define in the West as 'ethical' and 'political'. It is true that Buddhism has often been closely identified with political authority in many Asian countries, but as modern commentators have pointed out, it has shown little interest in developing a body of social and political theory of its own. It appears to lack a 'social gospel' and to be much less concerned with the struggle for social justice than either Christianity or Islam.[17] As Charles Wei-hsun Fu notes: 'In contrast to Christian tradition, the Buddhist tradition continues to lag behind in regard to the modern development of social ethics ... This, our first and foremost task, can no longer be evaded by the Buddhist community.'[18]

Religion and reproduction

Besides lacking a social gospel it may be thought that Buddhism is not well equipped to contribute to one specific area of bioethics, namely reproductive medicine. William LaFleur has recently drawn attention to the lack of interest shown by early Buddhism in fecundity and reproduction. He points out that its early literature has no place for the water-based myths of origin so common in other religions, and that it associates water with purity rather than fertility.[19] If anything, it was fire rather than water which became the emblem of early Buddhism. LaFleur comments: 'Fire sermons, a distaste for myths about fecund waters, a dissociation of *right* religion from anything having to do with sexuality and reproductivity – these were all constitutive of the Buddhism that is often thought to have been *original* or, at least, constitutive of the early stage.'[20] Another factor is that most Buddhist monks are celibate, and will therefore lack personal experience of the problems which arise in connection with reproduction and family life. Monks from traditional Buddhist countries may also have a different cultural perspective from that of the West on sexual relations, marriage and family life in general.[21]

There is little doubt that Buddhism has a lot of ground to make up if it wishes to address contemporary issues in a constructive way. If it is to flourish in the West (and all the signs are that it will) Buddhism must confront the issues which are on the agenda in the modern world. There is certain to be increasing pressure from Western Buddhists for clarification on many issues as scientific advances continue to be made, and the tradition will be forced to respond by adding a new discipline to its ancient curriculum. Buddhism has begun to make a contribution to Western civilisation through its profound philosophical and religious teachings. It is to be hoped that the West will in return make a contribution to Buddhism by lending its expertise in philosophical ethics.

III DOING BUDDHIST ETHICS

If we are to discuss Buddhist ethics at all, the first question which presents itself is how we are to determine what counts as a 'Buddhist view'. Buddhism is an ancient tradition with many branches, and one which has influenced and been influenced by the numerous cultures with which it has come into contact. The three most important forms of Buddhism encountered today are Theravada (the oldest surviving school of Buddhism), Tibetan Buddhism and far-Eastern Buddhism (including Pure Land, Zen and other sects). The historical spheres of influence of these have been as follows. Theravada Buddhism has been and remains prominent in South East Asia, notably in Sri Lanka, Thailand and Burma. Tibetan Buddhism influenced much of medieval central Asia but in modern times has largely been displaced from its homeland by the Chinese who have done their utmost to destroy Tibetan culture. Buddhism has a history of almost two thousand years in China, during which time its fortunes have ebbed and flowed. The situation of Buddhism in China today is difficult to determine, but there is no doubt that it has suffered greatly under communism.

Is there a 'Buddhist view'?

In view of this variety and the absence of any central authority, it is
difficult to make authoritative statements of the kind 'The Buddhist
view on issue x is ...' without qualification. It would of course be
equally misleading to make general statements about Christian
views without making clear whether the reference was to Orthodox,
Catholic, Protestant or other denominations. Despite the differences
amongst schools, however, it does make sense to speak of a
'Buddhist view' at least as far as our present purposes are concerned.
There is a great deal of consistency amongst the major schools in the
field of ethics, both in terms of the dominant pattern of reasoning
employed and in the conclusions reached on specific issues. Indeed,
there are good reasons for regarding ethics (particularly monastic
ethics) as a more cohesive force in Buddhism than doctrine.
Paul Williams has suggested that 'What unifying element there is in
Buddhism, Mahayana and non-Mahayana, is provided by the monks
and their adherence to the monastic rule.'[22] 'Thus', he concludes, 'in
spite of the considerable diversity in Buddhism there is a relative
unity and stability in the moral code.'[23] Theravada Buddhism,
Tibetan Buddhism and the majority of far-Eastern schools show
themselves in agreement as regards basic rules of conduct. Alternative perspectives are also found, but these minority views are
insufficiently representative to threaten the overall consensus. In the
light of this we can generalise with much greater confidence about
Buddhist ethics than we could about Buddhist doctrine. If it is legitimate to speak of a 'Buddhist view' on ethics, then, what is meant by
this phrase in the present context? What we shall refer to as the
'Buddhist view' is our own understanding of how the mainstream
tradition would, consistent with its underlying principles, begin to
formulate its reply to the challenges of modern life.

'The mainstream tradition' is used here to refer to the common
moral core which can be extracted from the different movements,
schools and sects. This core is is composed of common precepts,
values, beliefs and practices. We would expect the 'Buddhist view'
to represent the consensus among the majority of major Buddhist
schools and also to have a scriptural basis. To formulate this more
systematically, for a view to be described as the 'Buddhist view'
with the implication that it is orthodox or widely held, we would
expect to find : (i) authority for it in canonical sources; (ii) confirmation of it in non-canonical or commentarial literature; (iii) the

absence of contradictory evidence or counterexamples in these first two groups of sources; (iv) evidence that the view is pan-Buddhist (held by a majority of Mahayana and non-Mahayana schools); (v) evidence that the view is held across a broad cultural base; and (vi) evidence that the view has been held consistently over a long period of time. The more of the above points there are in favour of a view the greater the case for regarding it as an authentic expression of Buddhist principles.

The role of scripture

Reference was made above to the role of scripture in the validation of opinions which are proposed as candidates for the 'Buddhist view'. What are the grounds for this requirement? In the ancient debates a person who introduced a new opinion would be challenged by his opponent to 'bring the *sutta*' − in other words produce a text which supported his opinion. This approach to testing an opinion was commended by the Buddha himself, and we are told that in his last days he laid down conformity with scripture as the acid test for determining the validity of any opinion on religious matters. He counsels that if a monk volunteers an opinion it should be treated in the following manner:

> Monks, what is stated by that monk should neither be praised nor scorned. Without praise and without scorn every word and sylla-ble should be carefully noted and compared with the Discourses and the Monastic Rule. If, when so compared, they do not conform to the Discourses and the Monastic Rule you may conclude with confidence: 'Undoubtedly, this is not the word of the Lord, and has been wrongly grasped by this monk.' Therefore you should reject it. But if they conform to the Discourses and the Monastic Rule you may conclude with confidence: 'Undoubtedly, this is the word of the Lord, and has been rightly grasped by this monk.'[24]

It should not be thought from this that testing an opinion is simply a matter of establishing conformity with the letter of a text. A basic principle of Buddhist hermeneutics is that the letter must always give way to the spirit. Later sources often make a point of including a third test for validation in addition to the two mentioned by the Buddha. This third test clearly shows that it is not the text itself that is of importance but the truth it contains. The requirement is that whatever is stated in the text 'does not contradict the nature of things'.[25] This can be understood as meaning that the view expressed must not run counter to Dharma or natural law. The tradition also recognises that texts are of different kinds: in some the meaning is explicit while in others it is implicit and in need of interpretation.[26]

The four authorities

The above strategy for testing an opinion may work successfully in cases where there is a scriptural precedent for the matter at hand. But where does one turn for guidance if the question is not addressed directly in any canonical source? This problem was felt most keenly in connection with monastic discipline, and in his commentary on the Monastic Rule, Buddhaghosa formulates a more comprehensive hermeneutical strategy which builds on the above principle but also makes allowance for those cases where scripture is silent. He sets out this strategy in a formula of four authorities which are to be appealed to in order of priority until the point is resolved. These are:

1.   Scripture (*sutta*) itself
2.   that which is 'in conformity with scripture' (*suttanuloma*)
3.   the commentarial tradition (*acariyavada*)
4.   personal opinion (*attanomati*)

By the first, Buddhaghosa understands the Monastic Rule:[27] this is the final court of appeal in disciplinary matters. The second is a reference to a principle laid down in the Monastic Rule for dealing with matters which are not specifically prohibited.[28] The principle is that any conduct not explicitly ruled illicit should be regarded as prohibited if it is 'in conformity with what is improper (*akappiyam anulometi*) and opposed to what is proper (*kappiyam patibahati*)'.[29] The third is the commentarial tradition itself, which was thought to date back to the time of the First Council. As mentioned in the Introduction, this was a gathering at which the canon was traditionally thought to have been established and which is reputed to have taken place shortly after the Buddha's death around 404 BC.[30] Fourth and finally comes one's own opinion, which means much more than simple preference or unexamined sentiment. Personal opinion is

defined as 'the resolution of the question through logic (*naya*), intuition (*anubuddhi*), and inference (*anumana*) independently of scripture, what is in conformity with scripture, or the commentarial
tradition.'[31] Despite the requirement for careful reasoning,
Buddhaghosa specifically warns that too much reliance should not
be placed on conclusions reached even after such analysis. He further stipulates that the reasoning (*karana*) itself must be carefully
examined and its import (*attha*) checked against the scriptures. The
conclusion must also be checked against the two remaining superior
authorities (items 2 and 3) and rejected if not in conformity with
them, for as he points out, 'one's own opinion is the weakest authority
of all (*sabbadubbala*)'.[32] Buddhaghosa makes explicit the hierarchical
order of the four authorities:

> The commentarial tradition is weightier (*balavatara*) than personal
> opinion ... what is in conformity with scripture is weightier than
> the commentarial tradition. Scripture itself is weightier than what
> is in conformity with it, for scripture is incontrovertible. It is equal
> to the First Council in authority and is just as if the Buddha
> himself were alive today.[33]

It will be seen from the above that scripture is thought to play a
crucial validating role in questions of doctrine and ethics. It would
not seem unreasonable, then, to suggest that the credentials of a
view which is claimed as 'Buddhist' must be validated through the
application of the tradition's own standards. It follows that there are
reasonable grounds for scepticism towards opinions which describe
themselves as 'Buddhist' but do not make careful reference to the
relevant textual sources which the mainstream tradition holds
authoritative. This is not to say that all problems can be solved by
reference to scripture – far from it. It is simply to make the point
that scripture must be the touchstone for the assessment of opinion
which describes itself as 'Buddhist'.

Authentic moral conduct, it might be objected, lies not in slavish
obedience to the dictates of dusty manuscripts but in the exercise of
personal conscience. While the role of conscience in Buddhism
cannot be denied, the tradition holds that the proper exercise of
conscience yields conclusions which are consistent rather than
inconsistent with scriptural teachings. For Buddhism, scripture is
the embodiment of the Buddha's moral insight. The requirement for
conformity with scripture should therefore not be seen as the

idolisation of texts but as a check that one's own moral conscience is calibrated correctly. As we saw above, it is not the text itself that is important, but the fact that the text is 'in conformity with the nature of things'. One text illustrates this rather well when explaining what it is the texts themselves are to be checked against:

> With which Discourse (*sutta*) should the texts be collated? With the Four Noble Truths. With which Monastic Rule should they be compared? With the Monastic Rule (which combats) craving, hatred, and delusion. Against which doctrine should they be measured? Against the doctrine of Dependent Origination.[34]

The texts are thus a window through which the principles of natural law are discerned. It must be admitted, however, that the window sometimes needs a good deal of polishing before much can be seen through it.

Scripture in dialogue

The status of scripture within any religious tradition is a complex matter, and the two most common stances adopted with respect to it – on the one hand denying it any authority at all and on the other regarding it as a closed and exhaustive source of truth – are simplistic and misleading. Harold Coward has recently suggested that the relationship between a religious community and its scripture is best seen as reciprocal – a kind of dialogue within which the community defines itself over the course of time.[35] Buddhist literature certainly reflects changing attitudes and developments in the interpretation of doctrine. It has thus tended to function more as an 'open' than a 'closed' system. There are discourses in the Pali canon which post-date the Buddha's death, and the *sutras* of the Mahayana are the product of a literary tradition spanning many centuries.

Scripture is also 'open' in another sense, and postmodernism has drawn attention to the active role of the interpreter in the construction of meaning. Coward rightly reminds us that 'There is no escaping the fact that as scholars of Hindu and Buddhist texts we operate from within a hermeneutical circle.'[36] Rather than adopt the more radical conclusions of postmodernism, however, Buddhism would see scripture as both closed and open at the same time. It is open in that it remains a fertile source of new readings, but it is closed in the sense that not all readings are equally legitimate. Legitimate new readings will be those which progressively articulate what is implicit in the canonical sources. Opinions as to what is 'implicit' in the texts will be validated to the extent that these readings conform to the hermeneutical requirements set out by the tradition. In this way new interpretations will constitute a progressive unfolding of Dharma. Perhaps we could liken the creation of new meaning to a field which is generated between two poles, one of which is the text itself and the other the developing tradition in its encounter with new situations in the course of its historical evolution.

The above line of thought is of importance in two ways as far as our present interests are concerned. First, it means that in basing our arguments on scripture we are not moral archaeologists digging for fossils. Instead, we are raising questions which may never have been formulated before and demanding a response from the sources which is appropriate to the needs of today. The second point is that in emphasising the importance of scripture we do not see ourselves as engaged merely in the passive transmission of information. This is because our purposes, needs and aims themselves influence the selection and interpretation of the texts we employ. To reject scripture as irrelevant because it is 'out of date' is to fall into a naive objectivisation of the sources and miss the opportunity to generate creative new readings. The issues raised by Harold Coward will no doubt stimulate further reflection, but for now we turn from the problem of validating views as 'Buddhist' to the broader question of how Buddhism as an alien cultural tradition is to be studied in the West.

Cross-cultural ethics

Reaching sound ethical conclusions can be a difficult matter at the best of times, and engaging in cross-cultural ethics may be thought to be a methodological minefield. Some contend there are fundamental epistemological problems in understanding alien cultures at all, while others would allow the enquiry in principle but insist on a moratorium until basic rules of procedure can be established. Awareness of the problems in deciphering the moral languages of other cultures has been heightened by recognition of the increasing diversity and pluralism in the moral discourse of the West itself. Alasdair MacIntyre has likened the situation of the West to a civilisation in the aftermath of a nuclear catastrophe; the consensus which preceded the disaster has evaporated leaving groups of survivors clutching different fragments of the debris. More recently,

Jeffrey Stout has explored the problem of the diversity of morals using the metaphor of the tower of Babel. His quest is for a middle way between the Scylla of relativism and the Charybdis of a transcendent Moral Law. He sets out the basis for one possible path between the two whereby 'we are not left with any compelling threat to the possibility of moral judgement *per se* in cross-cultural settings'.[37] Stout's conclusion to this effect is positive in principle for our enquiry, and gives grounds for thinking that moral propositions can be universally true. He offers the statement 'slavery is evil' as an example. By this he means that slavery is wrong in any and all historical circumstances in which it might be found, regardless of local cultural circumstances. A corollary of this conclusion about the objectivity of morals judgements is the rejection of nihilism, scepticism and relativism. Although Buddhism agrees with Stout in rejecting these positions it does so, we suggest, for a different reason, namely that they are incompatible with natural law.[38]

Natural law

Stout does not accept that natural law can provide the 'universal, transcultural standard of morality' which he postulates as the alternative to relativism.[39] Stout's understanding of natural law as 'some transcendent thing in itself',[40] however, is quite different from that adopted here. His somewhat impoverished account of natural law (which conspicuously fails to mention the most important modern contribution to natural law theory)[41] presents it as a transcendent moral code against which earthly laws are to be measured. Quite rightly, he then rejects any such notion on the grounds that:

> You can't somehow leap out of culture and history altogether and gaze directly into the Moral Law, using it as a standard for judging the justification or truth of moral propositions, any more than you can gaze into the mind of God.[42]

Nothing quite so athletic (or egotistical), however, is envisaged by natural law theorists. Stout assumes that for moral truth to be objective there would have to be a back door out of culture giving direct access to the eternal 'Moral Law'. The precepts of a given culture could then be tested by popping out and comparing them against the Moral Law by simple visual inspection. Natural law ethics, however, has never claimed that moral truth can be established in this way. Contrary to common misperceptions, natural law reflection does not proceed by postulating transcendent realities, contemplating nature in the hope of deciphering the will of the Almighty, deducing moral laws from natural ones, analysing human nature in terms of its instincts, appetites and drives, or studying human societies in the anticipation of uncovering universal patterns of behaviour. Natural law does not maintain that objectivity in ethics has anything to do with a transcendent Moral Law in the sense this is understood by Stout. What it does claim is that objectivity in ethics is possible even though ethical reflection is always carried out by someone-or-other in some particular time and place. Natural law should not be thought of as a 'transcendent thing in itself' but a set of principles which guide reflection as to human good and the legitimate ways in which it should be pursued. In rather bald terms, natural law can be understood as asserting that there is:

(i) a set of basic practical principles which indicate the basic forms of human flourishing as goods to be pursued and realised and which in one way or another are used by everyone who considers what to do ... (ii) a set of basic methodological requirements of practical reasonableness ... which distinguish sound from unsound practical thinking ... thus enabling one to formulate (iii) a set of general moral standards.[43]

Natural law is concerned with the rational foundation of moral judgements. It begins with reflection on the basic forms of human good and ends with an account of which sorts of acts are reasonable all-things-considered in attaining these ends and which sorts are not. The principles of natural law are neither historical nor the monopoly of any one culture, and anyone reflecting rightly about human good will apprehend them in the same way. The objectivity of these principles is established by nothing other than the reasoning through which they are reached.

The historical evolution of *theories* about natural law in different cultures, on the other hand, is a proper subject for empirical enquiry. A history of natural law theories could be written in the same way that a history of scientific theories could. In the Introduction, we described the natural law tradition as 'a rope with many strands', and it is one to which contributions continue to be made. In the evolution of this tradition progress is made in fits and starts, often with long fallow periods. The contribution of Aquinas, for example, was a great advance on Aristotle in sketching out the particular ends which constitute human flourishing. At the same time there is much that is obscure, not to say mistaken, in Aquinas. In the period since Aquinas there has been much confusion and many dead-ends, but progress has been made in supplying the 'intermediate principles' (item (ii) above) which 'guide the transition from judgements about human goods to judgements about the right thing to do here and now'.[44] Clarifying and refining the principles of natural law takes place within a particular historical and cultural context, but the validity of the principles is not a function of their context. Buddhism, similarly, speaks of its own religious and ethical principles as eternally true regardless of the degree to which they are recognised in any time and place, or whether they are recognised at all. The Dharma is on occasion compared to a city in the jungle which exists despite all knowledge of it having been lost.

Since this is not a treatise on natural law we cannot pursue the theoretical aspects of the matter further at this point. The above rather compressed account should become clearer when we discuss the relationship between ethics and human good in Buddhism in Section V. For now, however, enough has been said to show that our understanding of the principles of natural law is quite different from Stout's. For this reason we believe it escapes his charge that natural law is deficient as a basis for cross-cultural ethics. On the contrary, we have found it to be an excellent analogue as far as Buddhism is concerned, and translation between the two systems is relatively unproblematic.[45]

Intercultural dialogue

Pellegrino rightly points out that ethics is not grounded by culture:

> The ethical system of any culture is morally defensible because it is grounded in truths that transcend that culture; it is not morally defensible simply because it is a product of a particular culture. Respect for culture and ethics other than our own is the beginning of any intercultural dialogue, not its ending.[46]

There are sufficient common denominators between Buddhist and Western thought to permit a fruitful intercultural dialogue, and to provide, in Robin Horton's terms, 'the cross-cultural voyager with his cross-cultural bridgehead.'[47] One important bridgehead is the ethics of medicine, and the fact that disease is a cultural universal[48] means that the ethics of medicine has a vital contribution to make to the dialogue. As Pellegrino puts it:

> As the biosphere expands to embrace the whole globe, every nation has a stake in every other nation's health. For these reasons, the practical and conceptual questions of transcultural biomedical ethics are more sharply defined than in some other domains of knowledge.[49]

Throughout this book ethical statements in Buddhist sources are treated as culturally transparent. We believe there are justifiable grounds for the inclusion of ethics in the enterprise which Paul Griffiths has described as 'cross-cultural philosophizing': this is an activity which presupposes the truth of the thesis that 'philosophy is a trans-cultural human activity, which in all essentials operates within the same conventions and by the same norms in all cultures.'[50] Buddhism presents itself to us as a set of doctrines and practices which are universal in scope, that is to say which are neither culture-specific nor historically determined. This is not to deny that Buddhism is itself an historical phenomenon which is always found in some specific cultural context or other, but only to point out that it sees its moral teachings as the expression of universal principles (Dharma). Rather than pursue a piecemeal enquiry into how Buddhists of school X in culture Y behaved at time Z, we propose to take Buddhist statements on ethics as philosophical statements which assert universal truths. While we are greatly interested in the *reasons* which lead Buddhism to its moral conclusions, it will be assumed, as in the sources, that valid moral conclusions can be universalised.

IV HUMANS, ANIMALS AND PERSONS

Central to many bioethical issues is the question of the nature and status of the moral subject. Some philosophers hold that a distinction can be made between moral subjects who are 'persons' and others who are not, and maintain that only 'persons' are entitled to full moral respect. Others reject this claim insisting that all *human beings* are worthy of full moral respect. Others again would extend the ambit of moral concern to animal and perhaps even plant life.

While most of the problems to be resolved in bioethics concern human beings, some attention must be paid to the fact that Buddhism adopts a wider moral horizon than is common in the West. Due to its belief in cross-species rebirth, respect for animal life is a prominent feature of Buddhist ethics. Respect for plant life is also in evidence, but there is considerable variation from school to school concerning the moral status of vegetation. Sources also vary in their views as to whether all forms of life are equally valuable, or whether the 'chain of being' is hierarchical. We cannot provide a comprehensive treatment of these matters here but will make some observations regarding the traditional Buddhist view of human and other forms of life, and sketch the outlines of a general approach to the question of the status of moral subjects which seems consistent with mainstream Buddhist opinion.

Human nature

We begin with a consideration of the Buddhist view of human nature, and it may be helpful if we compare this with the more familiar Christian model. Like Christianity, Buddhism sees man as a being with both a spiritual and a material side to his nature. Again, like Christianity, it regards human life as existing from the time these two elements are conjoined until the time they are put asunder at death. In the interim, man is faced with the challenge of realising the potential which his nature allows. Both traditions offer a role model as an example of the completed task: in Christianity the model for human perfection is Christ, in Buddhism it is the Buddha. For Buddhists the state of Buddhahood is the epitome of human perfection. The distance which separates us from this state is the ground we must traverse in our pilgrimage towards perfection, and Buddhism provides a structured path by means of which we can bring about the self-transformation needed if our spiritual profile is to match that of the paradigm. Whereas Christians may seek to emulate Christ but never in their earthly life achieve the perfection of his nature, Buddhists believe that every person has the potential to transform themselves fully in accordance with the example of the Buddha and to experience a state of perfection identical to that of the founder.

To bring the Buddhist concept of a moral subject into sharper focus, some understanding of the Buddhist view of human nature is required. As already noted, this consists of two parts; one spiritual, the other material. It is preferable to speak of these as two 'facets' or 'dimensions' rather than 'parts', for the latter term implies a dualistic view of human anthropology.[51] Although there is a clear sense in which the spiritual is separable from the material insofar as a material form is assumed at conception and abandoned at death, it would be more accurate to say that Buddhism sees man as a unitary being in a manner closer to the Aristotelian than the Platonic model.[52] As human beings we always exist simultaneously in the spiritual and material dimensions; even though our physical form may be different in another life, we will never exist as human beings without one. The spiritual aspect of human nature is therefore best thought of not as something separate and temporarily yoked to the body but as an aspect of the unitary being of the human individual.

The five categories

In recognising that man has both a spiritual (*nama*) and material (*rupa*) side to his nature the Buddha was not saying anything new in the context of Indian philosophy. He went on, however, to press the analysis further and to develop a new line of thought by listing five categories or dimensions in terms of which human nature can be analysed. This further analysis relates mainly to the spiritual (*nama*) side of the composite human reality. The canonical and commentarial explanations of the five categories are technical and complex, and what is provided in most summaries of this doctrine is a list of technical terms which do not greatly enhance understanding. This is perhaps due to the fact that the sources embark on their analysis with a specific purpose in mind: they do not make reference to the five categories as part of a neutral enquiry into human anthropology, but to illustrate their importance for soteriology. What follows is therefore an interpretative account rather than a restatement of the standard canonical definitions.[53]

The first and simplest of the five is *form*. Although not exactly equivalent to 'matter' this may be thought of as denoting the physical substance of the body. This physical dimension is, perhaps, the most basic and constant aspect of human experience. Buddhists understand the body to be created by one's parents and as having developed into its present form in accordance with the genetic programme established at conception. However, this process cannot be explained in material terms alone, and the presence of a spiritual element is required if matter is to evolve in a human form.

Buddhism believes that this spiritual nature pre-exists each life and is not, like the body, created by our parents. In terms of the categories of Western theology Buddhism therefore rejects the thesis of traducianism which claims that both body and soul are supplied by the parents.

The second of the five categories is *feeling,* and this denotes the capacity to respond affectively to a stimulus. Feelings are classified as pleasant, unpleasant, or neutral, and the most basic kind of feelings are simple sensations of the stimulus – response kind. An example of an unpleasant sensation might be to be pricked by a pin; a pleasant one would be a drink of cold water on a hot day. In addition to the capacity for *feeling,* human beings also have the power of *thought,* and this constitutes the third category. This includes the capacity to discern, discriminate and conceptualise, for example to name and distinguish different colours. The picture of man we have sketched so far is abstract and two-dimensional, and lacks any reference to the features which distinguish one person from another. These are the elements which constitute the fourth category.

Granted the power to think and feel, individual development will be shaped by personal experiences and reactions to them. From these reactions are built up particular tendencies, traits and habits, and eventually the complex pattern of dispositions which is referred to as *character.* It is the particular configuration of these traits and characteristics which defines people as the individuals they are. Commentators drew up long lists of virtues, vices, and other mental factors in order to provide an exhaustive account of this fourth category. Psychology here takes on a moral dimension, in so far as it is on the basis of their feelings and beliefs that individuals make choices, and these choices reinforce the pattern of their subsequent moral development. In essence, the fourth category denotes the patterns or 'complexes' of thought and feeling which have become habitual within a given subject.

Retrospectively, the fourth category is the culmination of a person's moral history. It is the sum of the moral choices made in the present and previous lives, and it will be instrumental in shaping the course of future moral development in this life and those to come. These long-term implications of character are what Buddhism means when it talks about 'karma'. The doctrine of karma might be summarised succinctly as the belief that a person's character is his destiny. From the point of view of ethics, the fourth category is of great importance. It explains how individuals shape themselves through the moral choices they make, and also confirms that they bear the final responsibility for the consequences of what they do.

The fifth category, *viññana*, is far from easy to describe and is not easily characterised. Discussion of its meaning is hampered by the need to resort to Western vocabulary which is saturated with potentially misleading associations. The usual translation of *viññana* is 'consciousness', but this is a term which itself abounds with difficulties. Nor is our predicament helped by the fact that the precise meaning of *viññana* is often coloured by the context in which it appears in the original sources. 'Consciousness' can be misleading as a translation since it is easily confused with the mental 'stream of consciousness'. The experience of *viññana* in this form, however, is merely one of its many modes. It is better understood as functioning at a deeper level and underlying all the powers of an organism. It is by virtue of *viññana* that we have bodily sensations, that we see, hear, taste, touch and think. *Viññana* resembles certain Aristotelian-derived notions of the soul in Christianity, namely as 'the spiritual principle in man which organises, sustains and activates his physical components'.[54] The term 'sentiency' is preferable to 'consciousness' since it is not restricted to the mental sphere in quite the same way. Sentiency can include bodily as well as mental sensations, and captures rather better the organismic sense of *viññana*. In the context of ethics, however, there is a source of possible misunderstanding in that 'sentiency' has itself been given a distinctive inflection by ethicists who seek to ground moral status in the capacity for suffering. In order to minimise confusion, therefore, the Pali term *viññana* will be retained in its original form.

While on the subject of terminology there is one further English term we should mention, although the context in which it is used will not concern us until the next chapter. Most writers on Buddhism avoid the word due to the particular associations it has with Christian doctrines on the soul. There are times, however, when the refusal to use the obvious English term hinders rather than helps the process of understanding. The term in question is 'spirit', and I do not think it would be misleading to refer to *viññana* in certain contexts as the *spirit* of an individual. *Viññana* is the spiritual DNA which defines a person as the individual they are. Following death it fuses with a new biological form giving rise to a being with a new physical body but an inherited moral profile. We might say that *viññana* provides the continuity of moral personality between a deceased person and the new genetic product. Accordingly, when reference is made to *viññana* as the carrier-wave of a person's moral identity, for example in the state of transition between one life and the next, it may be referred to as the 'spirit'. An alternative designation for *viññana* in the state of transition between lives is the *gandhabba*, which will be translated as the 'intermediate being'.

To sum up. In terms of the first three analytical categories it might be said that human beings are constituted by (1) a physical bodily organism which has the capacity to (2) feel and (3) think. The individual use made of these capacities leads to the formation of (4) particular habits and dispositions which distinguish each person as the individuals they are. Although feeling and thought define the architecture of experience, it is (5) *viññana* which constitutes it. It would be wrong to regard *viññana* as the *subject* of experience, as if it were a spectator peering out through the windows of the senses. Buddhism denies there is any such 'ghost in the machine' and maintains that *viññana* is dynamically involved in all experience whether physical or intellectual. Thus *viññana* arises in the form of vision, hearing, touching, smelling and tasting, when the eyes, ears, body, nose and tongue are in contact with their respective objects. The structure of the neocortex allows *viññana* to function in varied and complex intellectual modes such as reflexive self-awareness, memory, and imagination. As is common in Indian philosophy, Buddhism regards the mind (*manas*) as a sixth sense and not, as is presupposed in so much of Western philosophy, the very essence of a human being. In fact, Buddhism views the identification between the self and the intellect as a great obstacle to understanding the most important truth of all about human nature, namely that it has no enduring self or essence.

Humans and computers

The doctrine of the five categories is not easily grasped, and an analogy of the relationship between a computer and its components might help clarify their interrelationship. We must be careful not to press the comparison too far, however, for Buddhism does not think that human beings are merely complex machines. With this caveat in mind, we might compare the 'hardware' of the computer system to the human body. Thus the casing, components and circuitry of the computer are comparable to the physical substance of the human body. The 'software' or particular program being run by the computer corresponds to the fourth category in that a computer operates in accordance with a programme just as individuals operate in accordance with their characters. Of course, people have the freedom to change or 'reprogramme' themselves through the choices they make whereas computers (at least hitherto) do not. The fifth category, *viññana*, may be compared to the electricity which is needed to power the system. An electrical current flows through the computer and is invisibly present in every functional part. When the power is on, many complex operations can take place; when the power is off the computer is a sophisticated but useless pile of junk. Like electricity, *viññana* empowers an organism to perform its function. At the risk of pushing the analogy too far, the images which appear on the monitor screen might be likened to the stream of consciousness which flows through the mind. Death might be equated with the loss of the electric circuit due to the failure of a key component. Finally, the reinstatement of a person's moral character in a new existence might be likened to the transfer of data from one machine to another by means of a back-up copy.

Buddhism and 'persons'

We noted at the start of this section the view of some philosophers that 'personhood' should be the criterion of moral worth. On this view we must respect 'persons', but not all human beings or other forms of life. We may now enquire further what is it to be a 'person,' and what relevance, if any, this concept has for Buddhism. Philosophers disagree as to the precise criteria of personhood but most discussions of the subject take their cue from Locke, who defined a 'person' as follows:

> To find wherein *personal identity* consists, we must consider what *person* stands for; which, I think, is a thinking intelligent being that has reason and reflection and can consider itself as itself, the same thinking thing in different times and places; which it does only by that consciousness which is inseparable from thinking and, as it seems to me, essential to it; it being impossible for anyone to perceive without perceiving that he does perceive.[55]

A more recent definition of a 'person' is someone who is 'rational, is capable of free choices, and is a coherent, continuing and autonomous centre of sensations, experiences, emotions, volitions, and actions.'[56] Both these definitions of 'personhood' take the rational human adult as their paradigm. It follows that while all 'persons' are human beings, not all human beings are 'persons'. For example, the further we move back in the development of the individual human being the more difficult it is to be sure that these features are present. It appears that before birth many of the elements of personhood are lacking. It would be very difficult, for instance, to argue for self-consciousness in the early embryo before the development of the brain. Philosophers who apply Locke's views on personal identity to ethics therefore maintain that in the early stages of life there is only biological material which is at best a 'potential person' but which is not yet entitled to the moral respect reserved for 'actual persons'. They would also hold that an adult with severe dementia, such as advanced Alzheimer's disease, is no longer a 'person'.

Similarities and differences

Does the concept of a 'person' have any relevance for Buddhist ethics?[57] At first sight it appears that it might. The Buddhist view of human nature that we have set out readily encompasses the two definitions of a 'person' given above, and Buddhist psychology distinguishes all of the attributes (such as rationality and reflexive awareness) listed there. Furthermore, there is an historical parallel of a kind in that just as the Western notion of 'personhood' may be thought of as replacing the soul as the criterion of moral worth, the Buddhist doctrine of 'no-self' (*anatta*) was intended as an alternative to the Brahmanical 'self' (*atman*). Buddhism does not ground its ethics in a metaphysical soul or self, and denies that any such thing exists. According to Buddhism, the five categories are what remain when the 'soul' is deconstructed. The belief in a soul is a case of mistaken identity whereby the five categories are mistaken for a self. In view of this disinclination to seek a metaphysical basis for moral respect, Buddhism may be thought of as favourably disposed towards a more empirical one such as that provided by the concept of 'personhood'.

On the other hand, the particular features of 'personhood' mentioned above seem limited when placed alongside the Buddhist doctrine of the five categories.[58] Buddhism regards man as a complex of mental and material elements with a history and destiny which transcends a single lifetime. This nature is certainly not exhausted by the attributes of 'personhood'. For Buddhism, these represent the temporary flowering of certain capacities. The various features of 'personhood' are seen as arising naturally at the appropriate stage of biological development. These capacities, moreover, are fulfilled in degrees: they arise and disappear in a series and fluctuate even in a mature adult. The view that the moral worth of a human being arises and disappears as these capacities come and go is an idea which finds no support in Buddhist sources. Buddhism would insist instead on the psychophysical totality of man as the only legitimate basis for the attribution of moral status.

'Persons' and rebirth

The concept of 'personhood' appears increasingly irrelevant when viewed in the context of the Buddhist belief in rebirth. Buddhism understands individual existence as a continuum with a long trajectory. Within this trajectory an individual life may manifest itself in different forms at different times, and in any one life there will be development through various stages. It may help to represent these in the form of letters, where A stands for conception and Z for death. In between, *B* could stand for life in the womb, C for childhood, P for the points when the requisites of 'personhood' are fulfilled, and X for the time when they are lost after having been present, for example by relapsing into an irreversible coma. An individual life could thus be represented by a series of letters, with a normal lifespan expressed in the form *ABCPZ*. Our code can now be used to depict a series of four human existences which we will attribute to 'William'. In his first life William died in childhood (*ABCZ*), in his second he lived a normal life (*ABCPZ*), in his third he became irreversibly comatose (*ABCPXZ*) and in his fourth he died in the womb (*AZ*). The letter P features in two of these lives, which means that William was only a 'person' in half of his human existences. In the other two he was therefore something else, presumably, a 'non-person'. Because of this, it would have been legitimate to treat William in radically different ways at different times. When a P appeared in the code his rights would have been respected, while at other times he would have been treated as a non-person. This change in moral status would not be due to anything William himself had done and would be totally beyond his control: others would have determined under what circumstances he was to be treated with moral respect. There is an obvious danger here in that the indicators for 'personhood' could reflect the interests of those defining them. In this connection Alan Donogon has described the

contemporary concept of the 'person' as 'a do-it-yourself kit for con-
structing a "moral community" to your own taste.'[59]

Buddhism would reject the notion that an individual is a moral
being at certain times but not at others. It takes the view that all of
the letters used above are episodes in William's biography, and that
his moral status throughout remains unchanged. The extended
biography of any individual can be represented by a long string of
such letters, and if we run them together it becomes clear that what
is fundamental is not the individual letters but the continuity of the
series. Thus the code *ABCZABCPZABCPXZAZ* represents one and
the same subject over four lifetimes. From a moral perspective the
changes are superficial, and have no bearing on the status of the
subject. The sequence *PX*, for example, does not mean that William
ceased to be William and became something else (a vegetable?); nor
does the occurrence of letters other than *P* mean that William was
only 'potentially' William on these occasions, while being 'fully'
William on those occasions when a *P* appears. What all of the letters
mark are individual stages in the constantly shifting pattern that is
individual existence. What determines respect for William as far as
Buddhism is concerned is not any one letter in the code but the code
itself. He is entitled to moral respect because he exists as a living
being with a continuous biography and a spiritual destiny.

No-self

Buddhist doctrine places a further difficulty in the way of selecting
particular abilities as the ones to bear the moral weight of 'personhood'. The conclusion of the Buddha's analysis was that there was
literally *nothing* which could be thought of as the pith or essence of
human nature. The Buddhist denial of a self means that no *one*
factor from the total physical and psychological complex can be
singled out as more or less 'essential'. If no one factor can be singled
out in this way, the clustering of any two or three has an arbitrary
look about it.

There remains the possibility that although 'personhood' cannot
be expressed in terms of the five categories either individually or
together, it could be a phenomenon which arises from the interaction between them. There was, in fact, an ancient heresy in
Buddhism which claimed precisely this. According to this view
something called a 'person' (*puggala*), a kind of pseudo-self, was
said to emerge in this way from the interaction of the five categories

as a group. The fact that this view was universally condemned, however, seems to rule out any hope of legitimating 'personhood' understood along these lines by reference to orthodox Buddhist doctrine.[60] While the everyday word for a person is common enough in both Pali (*puggala*) and Sanskrit (*pudgala*), it is not used in the artificially narrow sense we are discussing here. One early text defines the word 'person' (*puggala*) as simply 'the continuous existence of any given living being'.[61] Collins suggests that the Pali *puggala* denotes mainly 'character-types', and has to do with 'differences in character, ethical disposition, spiritual aptitude and achievement, and karmic destiny.'[62]

Our conclusion must be that the notion of an individual as a human being but not a moral person is one which is alien to Buddhist thought. It is important to realise that it is alien not because Buddhist psychology lacked the conceptual sophistication to make the distinction, but because it saw no reason to. Such a distinction is simply incompatible with Buddhism's holistic understanding of what it means to be human. Buddhist psychology analysed mental phenomena in great detail, and distinguished over fifty separate mental faculties. Various permutations of these could easily have been aggregated into a bundle and labelled 'personhood' if Buddhism had wished to do so. All of the elements found in modern definitions of personhood are readily at hand in its psychological taxonomy, but it evidently saw no reason to identify certain features as exclusive markers for moral worth. Nor is the concept of a 'person' alien because of the distinctive emphasis on the role of the individual in the cultural evolution of the West. Buddhism has always made the individual central to its ethics, and would reject the criterion of 'personhood' for the very same reasons it is rejected by the natural law tradition in the West, not because of cultural differences but because of cultural similarities.

'Persons' and animals

Another reason why the criterion of 'personhood' would be rejected by Buddhism is that it involves a narrowing of the moral universe whereas the Buddhist inclination is to expand it. The move from respect for human beings to respect for 'persons' results in the exclusion of not just some human beings but the animal kingdom as well. Some philosophers hold that certain animals, such as the higher primates, can be 'persons', but even on this definition the greater part of the animal kingdom would still be excluded. What Buddhism would seem to require is that the movement be in the opposite direction and that moral concern be bestowed more liberally rather than more narrowly.

An important implication of the doctrine of karma is that forms of life are interchangeable. It is held, for example, that over the course of time humans can be reborn as animals and vice versa. This belief has a profound effect on how Buddhists see the animal world and how they determine their moral responsibilities towards it. A quick perusal of Asoka's edicts on Dharma dating from around 250 BC reveals that around a third of them contain references, sometimes lengthy, to animal welfare.[63] The fundamental division between man and the rest of creation which has influenced Western thinking since Genesis finds no foothold in Buddhist thought. Unlike the Western tradition, Buddhist sources stress the community between humans and animals rather than highlighting the differences. To the Western mind being reborn at all is a difficult concept to come to terms with, and rebirth as an animal seems a possibility no less remote than it is bizarre. It may be that Western Buddhists will seek to reinterpret the doctrine metaphorically, perhaps by picturing animal existence as symbolic of the 'beastly' quality of life of one who lives wrongly.[64] However, the ancient texts are clear that the possibility of rebirth as an animal is to be taken quite literally.[65]

Sentiency

We can visualise the relative extent of the three moral domains we have discussed above – persons, humans, and animals – in the form of three concentric circles of increasing size. The smallest circle includes only persons; the middle circle includes both persons and human beings; the third circle embraces persons, human beings, and animals. Philosophers such as Bentham have suggested that we should focus on just one capacity – sentiency, or the capacity to feel pain – and make this the sole criterion of moral concern. Whatever can feel pain, it is suggested, has a claim upon us by virtue of this fact alone, and this gives all sentient life an interest in not being made to suffer.[66] Since the capacity to feel pain is clearly enjoyed by most members of the animal kingdom in addition to human beings, we have a basis for extending moral concern to all forms of sentient life. Our moral universe would accordingly consist simply of one large circle with no separate reference to either human beings, persons, or animals.

This argument has some plausibility in a Buddhist context in three ways. First, it grounds the respect for life for which Buddhism is renowned in a rational moral principle. Second, it has the merit of providing a means of demarcation whereby a line can be drawn to determine which parts of the natural world belong within the moral sphere. Third, the reference to pain strikes a chord with the concern which Buddhism shows for the reduction of suffering. We could go so far as to say that the whole institution we know as Buddhism is geared towards the elimination of suffering, and owes its very existence to the compassionate concern of the Buddha for the suffering of sentient beings.

Taken cumulatively, these factors create the impression that the principle of sentiency has an important role to play in Buddhist ethics. On closer examination, however, a number of problems arise. Looking at the three factors in reverse order, it has already been suggested that the reduction of suffering which Buddhism aims at means much more than the elimination of physical or psychological pain. Buddhism does not set out merely to offer an anaesthetic for life, but to provide a solution to the inherent unsatisfactoriness of any kind of embodied existence. This goes far beyond considerations of temporary discomfort. While a life free from pain may be good as far as it goes, it would not be seen by Buddhism as counting for very much in the context of its major aims. Such a benefit would most likely be explained as due to good karma, and interpreted as a by-product of the moral life rather than as central to it. When the capacity for pain is placed in this broader context it becomes difficult to accord it such a fundamental role.

The second point, concerning sentiency as a demarcation principle, may be thought to have rather more potential in a Buddhist context. Buddhist sources tend to be somewhat inconsistent in the way they relate ethics to the evolutionary order, particularly at its lower reaches. Buddhism allows moral status to animals, and often seems to extend this to insects and microbes. We read in the early sources, for example, that monks used water-strainers to avoid harming the tiny organisms that live in water.[67] They also took up settled residence in the rainy season, in part to avoid to avoid treading upon the tiny creatures which come to life after the rains.[68] Do these factors show that the Buddhist respect for life extends to the microscopic level?

There is reason to hesitate before drawing this conclusion. One problem is that it is difficult to be sure whether these practices were inspired by moral concern or driven by lay expectations in a competitive religious environment. Buddhist monks were dependent on the laity for alms and would not wish to appear less rigorous in their eyes than rival religious groups. Many monastic precepts came about directly as a result of complaints from the laity, and these complaints often explicitly compare the behaviour of Buddhist monks with that of rival mendicant groups. An example can be seen in the account given by the Monastic Rule of the circumstances in which the prohibition on travel during the rains came about.

> At that time the Blessed One had not yet imposed on the monks the rule regarding the Rainy Season Retreat; the monks travelled both during the summer and during the rainy season. People were annoyed and complained angrily: 'How is it that these ascetics, the sons of the Sakyans (i.e. Buddhists), keep on travelling during the summer, winter and also in the rainy season? They tread on young plants and damage them, and destroy many small living creatures. *Those who belong to other schools* may not be very well-disciplined, but at least they withdraw somewhere to make a residence for the rainy season.[69]

The above raises some doubt about the *moral* content of monastic prohibitions which arose in this way. Although it does not show conclusively that they have none, allowance must be made for the public-relations factor which is undoubtedly present. The laity would certainly compare the behaviour of Buddhist monks with that of their peers, notably the Jains, who are particularly scrupulous in such matters. Rather than lecture the laity on the intricacies of moral philosophy, the Buddha may have deemed it prudent to make modest concessions to their expectations. In this way harmony would be restored and the extra vigilance required of monks would be good training in mindfulness. It will be seen that the passage above makes reference not just to organic life but also vegetation, and monks are elsewhere cautioned not to cause damage to seeds and plants.[70] Perhaps again this is best seen as an issue which concerns monastic etiquette and deportment, and has more to do with the public image of Buddhism than its ethics. Although it would be unwise to draw final conclusions at this stage, details of this kind should not be seen as providing conclusive evidence that the Buddhist respect for life extends to vegetation.

Given the absence of any clear guidance in Buddhist sources as to where exactly to 'draw the line', sentiency would seem to have much to commend it. For example, it would exclude both vegetation and biological organisms such as amoebas and the like in which the capacity for pain is presumably absent. While there may be some grey areas these could be argued over and the facts of each case considered in turn. Overall, it gives us a useful rule of thumb which corresponds to the common-sense intuitions of most people who, for example, love their pets but do not lose sleep over the fate of their pets' fleas.

Problems with sentiency

The main attraction of sentiency as a moral criterion for Buddhism is that it is relatively unproblematic in its application to the animal world. But is this a sufficient reason why Buddhism should adopt it? While the capacity to feel pain might be a useful indicator of where the boundaries of the Buddhist moral universe lie, it does not follow that it is *because* of this capacity that forms of life deserve moral respect. Sentiency is one of the capacities of living organisms, but *only* one of them. There are times when it may seem to be the most important of all, as anyone who has suffered a painful toothache will know, but on more sober consideration we may doubt that it has such central importance.

The claim that sentiency should function as a moral *principle* rather than an *indicator*, gives rise to problems of the kind considered in relation to 'personhood'. When applied to humans, the choice of sentiency as the litmus test of moral worth would mean that the second of the five categories of human nature had been singled out for priority. It would be saying, in effect, that the Buddha should not have produced a list of *five* categories in his analysis of man since what is most essential in human life can be found in the second one, namely sensation (*vedana*). This suggestion, however, is specifically refuted by the Buddha,[71] and there is no evidence in Buddhist sources that any single one of the five categories is or should be viewed in this way. When the five categories are mentioned, they are usually talked about *en ensemble*. The Buddha's exposition of the doctrine of the five categories suggests that all five are co-ordinated aspects of one entity. If this is correct it follows that no *one* of them can capture the essence of a being, and since no *one* of them can, neither can any two or three, or any permutation less than five.

We might take this opportunity to point out that although we are discussing this possibility here with respect to sentiency, the same reasoning would apply to any of the other four categories. It might be thought that those related to the intellectual faculties (the third) or to intention and volition (the fourth) were better candidates for this purpose than sentiency. It might be also be thought, not unreasonably, that the most fundamental is consciousness (*viññana*), the fifth. To specify *viññana* as the criterion of moral status is, however, simply to say that all living beings have moral status, since it is impossible to isolate *viññana* from the psychosomatic totality of a living being. It is impossible to point to *viññana* without in the same act pointing to a living creature, just as it is impossible to point to 'shape' without referencing a physical object. Overall, since neither *viññana* nor any other of the five categories by themselves can adequately encompass the nature of a living being, there is reason to be suspicious of any view which claims to locate in any one of them what is essential in human nature.

Two further problems might be raised in connection with the attempt to ground moral status in sentiency. The first arises from the fact that it is possible to separate conceptually the power of sentiency from all the other powers of a being. It is possible, for example, to imagine a human being (or animal) which has for one reason or another lost the ability to feel pain. Such a being might, for example, be a genetic freak. Would it lack moral status? Again, what if scientists eliminated pain from the human species through genetic engineering? What would then be the basis for moral relationships? In a pain-free world there would be no logical foundation for ethics. The implausibility of ethics dying out along with the capacity for pain leads us to think that it would be a mistake to adopt this factor alone as the determinant of moral status.

The second problem is that if the primary moral injunction is not to cause pain, there could be no objection to killing painlessly. Imagine a case where, with no advance notice, a creature was killed painlessly during its sleep. Of what wrong would the killer be guilty? Evidently none, unless considerations of autonomy were brought in, but this would make the act wrong for reasons which have nothing to do with pain. The reasons why Buddhism values life do not seem to centre on sentiency, and Buddhist sources show little interest in a hunt for features of this kind which might be used to distinguish particular life-forms as moral beings. On the contrary, it treats them as moral beings simply by virtue of what they are, namely living creatures. The evidence from Buddhist sources suggests that living beings are worthy of respect simply by virtue of the inherent dignity which is inalienably theirs as living beings. In other words, for Buddhism, life has intrinsic worth. We will develop this point further in the next section.

V. ETHICS AND HUMAN GOOD

We must now enquire as to the principles by which Buddhism reaches its ethical decisions, for without some knowledge of them it will be difficult to understand the rationale which underlies scriptural precedents. Unfortunately, this is easier said than done, since the principles underlying Buddhist ethics are rarely made explicit in the sources. Although different formulations of moral precepts are constantly encountered, comparatively little effort is expended in providing a justification for them. A satisfactory justification would provide an account of the conceptual relationship between the precepts and the Buddhist vision of human good. In other words, it would explain the connection between ethics and enlightenment. To discover the form such a justification might take we must first review the main features of ethical choice and action.

The Buddhist moral landscape

We might begin by taking an overview of the moral terrain and noting the topographical features of this landscape which claim attention. One sees the high ground of motivation, the peaks of intention, the rugged terrain of action, and the rolling foothills of consequences disappearing in the distance. Faced with this panorama, where should attention be focused in order to isolate those factors which distinguish moral from immoral action? We can hardly keep the whole panorama in focus at once, and yet to single out any one aspect results in a loss of perspective overall. Nevertheless, we must begin our fieldwork somewhere, so let us explore the role of consequences first.

Consequences and karma

The pattern of validation which immediately suggests itself in Buddhism is a consequentialist one, that is to say, one which justifies the precepts by reference to the consequences which flow from keeping them. When asked why they keep the precepts Buddhists will typically make reference to the heavenly rebirth which is thought to be secured through good karma. This is hardly surprising, since Buddhist literature everywhere makes clear the connection between moral conduct and good consequences of this kind. Buddhists certainly believe that moral acts bear fruit, as do Christians, Jews and Muslims. When Christians are asked why they keep the commandments they, like Buddhists, will tend to reply by pointing out that good deeds will be rewarded in heaven. Responses of this kind deflect attention away from the acts themselves and onto their consequences, creating the impression that what is valued is something which is extrinsic to the deeds performed.

A more reflective analysis, however, will show that the goal of a heavenly rebirth in both Buddhism and Christianity is not to be reached by an arbitrary set of actions which fortuitously produce good results, but by an individual making themself *worthy* of the goal. Indeed, those who achieved realisation through following the Buddha's teachings are referred to as 'Worthy Ones' (*Arahat*), and it is only doing certain *sorts* of things which make a person worthy of the goal. Once this is recognised, the search for justification shifts away from consequences to the distinctive characteristics of the sorts of things which need to be done. While longer-term consequences are important in moral action, what is of greater importance for both Buddhists and Christians is the day-to-day making of their moral selves. Buddhism holds that if an individual makes the right moral choices, he will reap a reward both in the present and in the future; if he makes immoral choices he will suffer the consequences in both the short and long term.

It must be stressed that the concern about the effects of actions on oneself and others which is found in Buddhism does not make Buddhism 'consequentialist'.[72] Consequentialist theories of ethics (of which utilitarianism is an example) maintain that the rightness or wrongness of an act is determined purely by the consequences which flow from it. A simple consequentialist justification does not seem appropriate in the context of Buddhism in view of the emphasis it places on motivation, as we shall see below. This is not to say that Buddhism has no regard for consequences. There are many circumstances where it is intelligent to select one course of action from others purely on the basis of its likely consequences. This situation arises where there are several means to a good end, and none of the means under consideration involves an immoral act as part of it. In a situation of this kind it is clearly rational to prefer the course of action which will achieve the end most efficiently, maximise the benefits to all concerned, and minimise negative consequences. Although a consequentialist-like comparison of the likely outcome of competing alternatives is appropriate in cases of this kind, this does not mean that consequences determine rightness. On the contrary, Buddhism holds that certain acts are intrinsically wrong regardless of the consequences which flow from them.[73] If a consequentialist justification is rejected, however, some other explanation must be offered as to why the precepts prohibit certain kinds of conduct. Perhaps we will meet with greater success if we turn our attention from the results of actions to the motive from which they are performed.

The psychology of moral choice

When the Buddha defined karma he did so not in terms of consequences but by reference to moral psychology (*cetana*). The emphasis on the psychological factors which underlie moral choices was a deliberate one, and the Buddha was at pains to distance his teachings on moral conduct from certain of his contemporaries. His most prominent competitors were Brahmin priests, who seemed to the Buddha to be suffering from a spiritual sclerosis which distorted religious sentiment into stilted ritual. The obsession with complex and meticulous ritual observances whereby the mechanical performance of the act itself became all-important seemed to the Buddha to place far too much weight on externals. The neglect of the inner dimension of the moral life was also a feature of another contemporary movement, Jainism, which identified moral responsibility with overt physical action, regardless of the intent with which it was performed. Besides these, there were other contemporary teachings including various varieties of determinism, materialism, and scepticism, which failed to give due weight to the moral will.

To some extent, then, the Buddhist emphasis on the psychology of moral behaviour as distinct from its more overt aspects can be understood as a reaction against contemporary views. However, although the Buddha allowed motivational factors a privileged status in moral judgements, he seemed to think there was more to morality than simply 'meaning well'. Once, when asked for the criteria by which moral judgements should be reached, he made reference to objective factors such as consequences and the 'opinion of the wise'.[74] The reference to consequences was a reminder to think carefully about the effect that actions have on others. The 'opinion of the wise' involves more than just a majority vote among the learned; it implies that the contemplated course of action should find approval among those in a position to subject it to reasoned analysis in the proper Buddhist fashion using the criteria mentioned earlier. This requirement that both consequences and the 'opinion of the wise' should be considered suggests that the rightness of actions turns upon more than just the motive from which they are done.

It looks from our discussion as if neither of the two aspects of moral behaviour we have considered so far – the consequences of actions and the motive from which they are performed – are adequate by themselves to explain why certain things are right and wrong, although it would seem they each provide part of the explanation. Buddhism seems reluctant to narrow its perspective to these or any other single feature of the moral landscape. This is not because it cannot decide which feature is important, but because it believes that they *all* are. The dominant pattern of moral validation in Buddhism takes into account the psychology of the actor, the nature of the act, and the consequences which flow from it. Just like the landscape, all the principal features of a moral act from its motivation through to its performance and consequences, are inextricably interconnected. What is required, then, is a justification of the precepts which gives due weight to all of these factors.

We will suggest a way of approaching the question which does not base itself on explanations which are extant in the sources but which nevertheless seems to encompass all of the factors we have considered. This account will suggest that certain acts are wrong not because of their consequences, the motive from which they are performed, or the fact that they breach the precepts. Instead they are wrong because they are contrary to Buddhist values; to put it another way, they are in conflict with Dharma or natural law. To understand how and why immoral actions are in conflict with natural law will require a brief discussion of the relationship between the precepts and Buddhist values.

Precepts and goods

It has been suggested that the structure of an ethical system may be thought of as having four main levels.[75] On the ground floor are the things which constitute the everyday business of ethics, namely judgements, choices and actions. On the next level are the rules and precepts which justify the judgements and actions. Higher still are the principles which inform the rules, and finally above the principles is the ethical theory in terms of which the principles are to be defended. In terms of this model of an ethical system, Buddhist literature gives us access to only the bottom two levels. While there are many examples of judgements and actions, and various formulations of rules and precepts, we are thrown back on our own resources when we attempt to discern the content of the upper two storeys.

The task of determining the principles which justify the rules (the content of level three) is an exercise in reverse deduction. This is a sort of ethical algebra in which the outcome is known (the precepts) but the value of the other term in the equation has yet to be determined (the principles they defend). Given our present state of knowledge, what these principles might be is a matter of conjecture. The same data (such as judgements and precepts) can often be explained by more than one theory, and a number of ethical theories could present themselves as candidates to explain the data in the sources. We have suggested elsewhere that the most appropriate ethical theory in terms of which Buddhism might be understood is Aristotelian in form,[76] and this hypothesis, in conjunction with the principles of natural law, will now be used to explore the third level of the Buddhist ethical structure.[77]

The role of the precepts

The Buddhist precepts are couched in the form of an undertaking to abstain from certain kinds of acts, one of the most serious of which is the taking of life. It can be seen with a little reflection that the precepts gesture beyond themselves in the direction of certain values which it is their function to preserve. Their formulation as negative injunctions flashes an alert that anyone contemplating such actions as killing or stealing is threatening an assault on certain values or 'goods'. The precepts themselves can be fashioned in a variety of ways and can be rearranged and reformulated according to specific needs and requirements. Buddhism has various sets of precepts, the most common of which are the lists of five, eight, and ten. There are precepts for laymen, precepts for novices, precepts for monks, precepts for nuns, precepts for weekdays and precepts for holy days. Some of these are moral precepts, some are designed to inculcate restraint and self-discipline, and some regulate communal life. A truly exhaustive list of precepts would be one which detailed every conceivable circumstance in which a value was at risk. Such an undertaking, however, would be a futile scholastic labour, for the complexity of life would ensure that it could never be completed. The precepts therefore content themselves instead with flagging the most common ways in which important values may be threatened.

Seen in this way the Buddhist precepts are a set of normative rules which are derived from higher (logically precedent) principles which in turn seek to protect and promote certain values or goods. How can we know what these values might be? It is important to realise that the values we are seeking are not in themselves *moral* values, and that moral questions arise only when choices are made with respect to them. It may be presumed, for example, that readers of this book value knowledge. This affirmative valuation of knowledge does not in itself have moral implications. It is only when *choices* are made with respect to it that moral questions arise. For example, if the reader is an impoverished student presently browsing in a bookshop, a moral choice might arise about whether or not to steal the book in order to acquire (one hopes) knowledge of Buddhist ethics from it. Because the basic values we are seeking will not have an overtly moral dimension to them, we should not expect to find them linked directly to individual precepts. It may be more helpful, therefore, if we turn our attention away from the precepts for a moment and adopt a different tack by asking about the basic values of Buddhism as a religion.

Three basic goods

What vision of human good does Buddhism hold out? We may suppose that the purpose of the religious life is to realise certain goods. What might they be? To put it slightly differently: what will a person secure through following Buddhist teachings that he or she presently lacks? When the question is formulated in these terms the answer is not difficult to discern. Everywhere, Buddhism enjoins its followers to pursue two particular ends, namely knowledge and an harmonious moral relationship with others. In Buddhist terminology these things are usually spoken of as Wisdom and Compassion, and the prominence given to them by the tradition suggests that both must figure in some form or other in the cluster of core

Buddhist values. To these we will add only one more, one which is central to our current enquiry. We are alerted to the importance of this third value not so much by traditional formulations of worthwhile ends as by the constant emphasis upon its preservation found in the precepts and elsewhere. This third basic value is none other than life itself. From here on, these three fundamental values will be designated as *basic goods*, using the terminology of *life, knowledge* and *friendship*.

To say that life, knowledge and friendship are *basic* is to say that they are irreducible: for example, knowledge is not friendship, friendship is not life, and life is not knowledge. To say that life, knowledge and friendship are *good* is to say that these are the things which make for a fulfilled life as a human being. They are fundamental aspects of human fulfilment or flourishing in that each makes a unique contribution to the nature of the being one wishes to become (a Buddha). That they are good needs no demonstration since it is clear, for example, that knowledge is preferable to ignorance, and that it is better to have friends than to be lonely. Because life, knowledge and friendship are good they are also intrinsically desirable. They are not just instrumental goods in the way that, for example, money is. Money is good only as a means to obtain something else, whereas the basic goods are good in themselves. The three items listed do not correspond, as a group, with any standard Buddhist formula, but on reflection it can be seen that they are fundamental Buddhist values. We will discuss them together and then individually.

The basic goods in Buddhism

Knowledge figures as a good to be pursued in every formulation of the goals of the Buddhist religious life. The aims of early Buddhism can be seen from the Eightfold Path, where knowledge features as *Pañña*. In the Mahayana there is constant emphasis on the goal of knowledge as *Prajña*. There are fewer explicit references to friendship in the way the term is used here, although it does occur quite frequently in the more restricted sense of spiritual friendship (*kalyanamittata*).[78] What is designated here as the good of friendship, however, embraces a wider complex of ideas which Buddhism itself seems to have had some difficulty in labelling consistently. The essential notion is that of the proper mode of relationship with others, which in the West we would regard as the domain of ethics.

This is the complex which in early Buddhism is labelled as Morality (*sila*), and in the Mahayana as Compassion (*karuna*) or alternatively as Means (*upaya*).

While knowledge and friendship feature in the main soteriological registers of Buddhism, the third of our three basic goods – life – does not. This omission is not due to the fact that life is in any sense less fundamental a good than the others – indeed it is basic to them all – but that it is implicit in any formulation of human good. To pursue knowledge, friendship or anything else, one must first of all exist as a living being. Life is thus both a good in itself and a precondition for the fulfilment of other goods.

It is not claimed that this particular formulation of basic goods is comprehensive or exhaustive, but only that it at least overlaps the most important Buddhist values as found in the sources. There may be different and better ways of mapping these values, and it may be there are other human goods which Buddhism for one reason or another makes little reference to: aesthetic experience may be a case in point. All that is suggested so far is that Buddhism holds these things to be good in themselves and simultaneously legitimate ends or opportunities to be pursued and realised by individuals in their lives. It is through the pursuit of these goods that individuals progressively transcend limitations such as ignorance and selfishness and come to be more fully what their nature allows. If we take the Buddha as our paradigm, we see an individual who lived a *life* in which *knowledge* and *friendship* were pursued to the highest degree attainable in human terms. Many schools of Buddhism hold the view that the Buddha continues to enjoy a mode of life free of all imperfection in which knowledge and friendship are experienced on a macrocosmic scale. We will return to this point below, but for now let us say a little more about each of the three basic goods in turn, beginning with the good of life.

Life

One aspect of Buddhist ethics which strikes all commentators is its profound respect for life. The Buddhist respect for life is enshrined in the principle of non-injury (*ahimsa*), a principle so central to Buddhism that one text can claim 'Non-injury is the distinguishing mark of Dhamma.'[79] This means that non-injury, and the respect for life it presupposes, lie at the very heart of Buddhist teachings. Since Buddhism views its teachings as grounded in natural law it means that Buddhism holds respect for life to be a universal moral impera-tive. This imperative is given expression in the various sets of precepts, all of which include a prohibition on the destruction of life. The emphasis upon respect for life enshrined in Buddhist precepts, along with the effort expended in treating sickness by means of the healing arts, leads us to conclude that Buddhism shares the respect for life which is fundamental to the moral and legal traditions of the West. This belief in the 'sanctity of life'[80] should not be understood as a commitment to 'vitalism' (the belief that life must be preserved at all costs) but as the notion that intentional killing always repre-sents a failure to respect the inalienable dignity of living creatures. Granted that Buddhism places a high value on life, two important and interrelated questions arise: first, what is meant by 'life' in a Buddhist context; and second, are all forms of life valued equally?

Life and life's worth

The above questions are not explored with any rigour in the early sources, and there is no definitive statement as to which forms of life are valued and why. It has been rightly observed that life is 'a phenomenon having varied forms and disputed boundaries', and it is also a property which may be predicated of 'cells, tissues, organs and organisms, of plants, animals, humans and gods, of individuals, groups, species and systems.'[81] Some, like the Jains and Presocratics, would go further and see the whole of the natural world including inanimate matter as 'ensouled' or permeated with life. There are traces of this view in later, particularly far-Eastern, schools of Buddhism, which come close to adopting a Schweitzerian 'rever-ence for life' whereby plants, micro-organisms, and even natural phenomena are given moral status. Indo-Tibetan schools, on the other hand, tend to see the relationship between plant, animal and human life as hierarchical rather than egalitarian. Human life is regarded as the most auspicious of all rebirths and occupies a place at the top of the hierarchy.

Although Buddhist sources do not explicitly draw this conclu-sion, there would appear to be grounds for distinguishing two main categories of life. The two categories are those forms of life which can attain nirvana and those which cannot. Nirvana is the end of cyclic (*samsaric*) rebirth: as such, it seems reasonable to suppose that nirvana can only be attained by forms of life which are subject to rebirth in the first place. This would exclude certain of the examples of 'life' mentioned above such as 'cells, tissues, organs and organisms and plants' since none of these undergo rebirth. A lettuce will not be reborn, and it is difficult to conceive of it having an end or *telos* beyond its present existence. The same is true of bacteria and also of natural features of the ecosystem such as rivers, mountains, clouds and the like. Even if these things are 'alive' in anything more than a metaphorical sense, it is difficult to conceive of them having nirvana as their destiny.

Karmic life

The above suggests that the Buddhist respect for life is rather more nuanced than at first supposed. What is being suggested is that life has intrinsic value only when it possesses the capacity to attain nirvana. According to Buddhist doctrine only karmic life has this potential. To say that life has 'intrinsic' value is to say that it is affirmatively valued for its own sake rather than as a means to something else (i.e. its value is not instrumental).[82] By 'karmic life' is meant life with a karmic history, that is to say, life which has a moral biography. Karmic life can be contrasted with other forms of 'life' which do not undergo rebirth; since these have no moral past or present it is hard to see how they can have nirvana as their future goal or *telos*. If this conclusion is correct, it means that what Buddhism values is not merely life itself − life's 'livingness' − seen, for example, in vegetation, or in biological processes such as the growth of nails and hair, but life which has a spiritual *telos*.

Precisely which forms of life (which species) are telic is difficult to state with precision. Human life certainly is, and so, apparently, are the higher mammals. Traditional Buddhist paintings of the 'wheel of life' representing the *samsaric* cycle populate the realm of animal rebirth with creatures such as cattle, horses, dogs, tigers, birds, fowl and fish, and also more exotic mythological creatures. We may wonder what it is about these creatures that makes them karmic. Unfortunately, this is a question the sources do not explicitly address. The focus, as always in Buddhism, is on the practical question of how karmic life is to *fulfil* its *telos*, not the more speculative one raised above. It would seem, however, that at least two criteria must be fulfilled for life to be karmic: first, it must be sentient, and second, it must be individual. Reference was made earlier to the use of sentiency as a rule of thumb for determining the boundaries of the moral universe. Although the boundary is ragged at the edges, sentiency seems to provide one common denominator of karmic life. The second requirement is that the form of life be individual. The creatures referred to above as examples of karmic life share a common feature in that they are all identifiable as what might be called 'ontological individuals'. The concept of an ontological individual may be defined as follows:

> An ontological individual is a distinct being that is not an aggregate of smaller things nor merely a part of a greater whole. Although the millions of cells in our bodies are genetically identical, each one is not an ontological individual or separate entity. There is only one human individual that really exists in the primary sense of actual existence, though there are many cells that share in the existence of that single living ontological individual.[83]

Thus cattle, horses, dogs, tigers, birds, fowl and fish are ontological individuals in a way that the cells, bacteria and micro-organisms in their bodies are not. An ontological individual retains its identity through various stages of development whereas the cells of which it is composed may change. Thus William the adult is the same ontological individual as William the child, even though every cell in his body may have changed. Existence as an ontological individual, however, is not a *sufficient* condition for life to be karmic. The criterion would exclude cells, for example, but not a lettuce; a lettuce is an ontological individual but one without a karmic past or future.

A further point is that although all forms of karmic life have intrinsic worth it need not follow that they are equal in value. We noted above that animal life is not regarded by most schools of Buddhism as equal in value to human life. What is it that differentiates the various forms of karmic life? The answer seems to lie in what we might term their relative proximity to nirvana. The different capacities of species means that their scope for participation in nirvanic goods varies. Some, like man, can participate fully in the goods of friendship and knowledge while others, like animals, are restricted in the degree to which they can share in these things. Although the remarkable capacities of certain animal species are now well documented, few, if any, would be equipped, for example, to apprehend the truths of Buddhist doctrine intellectually. We should note that participation in nirvanic goods is not to be identified with their instantiation at any given time: it is not a question of whether one is reflecting on the First Noble Truth or being a friend *now*. If this were the case, it might be suggested that an adult dolphin or chimpanzee has more scope for participation in knowledge and friendship than a human embryo, and is therefore more valuable. The scope for nirvanic participation is determined instead by the extent to which one ha͟s *already* actualised the basic goods. A creature's physical nature is, according to Buddhism, a manifestation of its moral status. We might say that human nature is itself the product of moral evolution. No animal, therefore, can be more valuable than a human being, however intelligent the animal or however handicapped the human being. We will touch on this question again in Chapter 2 when we consider the reasons why the gravity of offences is thought to vary relative to the status of the victim.

Non-karmic life

A problem for the suggestion that only karmic life possesses intrinsic moral worth arises from a fact to which reference was made earlier. We noted above that early sources advised the use of strainers to filter water, so that monks would not destroy the tiny organisms it contains when drinking. Although we cannot be entirely sure, it is unlikely that these tiny organisms would be regarded by Buddhism as karmic life. Why, then, should precautions be taken against killing them? We suggest there are two possible explanations for this. It was suggested above that this concern can be explained by the need for Buddhist monks to be at least as scrupulous as their religious competitors. That the requirement to filter water applies only to monks and not laymen lends support to the supposition that the practice has more to do with protocol than ethics. To this can now be added a second explanation: this is that while non-karmic life may lack intrinsic value, it may nonetheless retain instrumental value. Its instrumental value lies in the fact that it provides the environment needed for karmic life to fulfil its *telos*. Tiny organisms and other forms of non-karmic life, such as vegetable and plant life, play an important role in maintaining the ecological balance which is required if human and animal life is to flourish. To destroy these forms of life is to threaten the well-being of the karmic community, and thus has serious moral implications. Vegetable and plant life is also of value in more direct instrumental ways, for example as a source of nutrition and oxygen. The basis for ecological concern in Buddhism is therefore to be found in the important role which the natural environment plays in sustaining karmic life. Since the ecology of the planet is a complex and delicate system, it follows that the natural environment should be treated with respect. The Buddha is represented in the early sources as something of a 'nature lover' with a predilection for simple natural environments.[84] Such environments were no doubt favoured because they were conducive to tranquillity and meditation. In other words, they were instrumental in furthering the participation of karmic life in its spiritual *telos*.

We may summarise our conclusions about the value of life in Buddhism by saying that (1) All karmic life has intrinsic worth; (2) Only sentient, individual life is karmic; (3) Not all karmic life is of equal value; (4) Non-karmic life has only instrumental value.

Life as suffering

Although we have described life as a 'good' it might be suggested that, on the contrary, Buddhism regards life as a *bad* thing rather than a good thing. Does it not, after all, constantly draw attention to the suffering inherent in existence of any kind? Here two points should be made. The first is that the negative statements one finds in Buddhist sources about life are usually in the context of life when it is lived wrongly. A great deal of suffering is self-inflicted and comes about through living in conflict with Buddhist teachings rather than in harmony with them. To counterbalance this, the sources extol the joy and happiness to be found in a life lived rightly. On the philosophical side, we must say that when Buddhism points to the inherent unsatisfactoriness of life in *samsara* it is not saying that life as we now know it is not good, but only that it is *less* good than the more perfect form of life attained in final nirvana. A Christian may make the same distinction between this life and the one to come. The fact that there exists a transcendent backdrop to our present existence does not devalue the life we now enjoy. It is unclear from Buddhist sources what form life in nirvana will take, but it will take *some* form, since Buddhist doctrine condemns as heresy the view that nirvana is annihilation. Indeed, Buddhism holds that karmic life *already* enjoys eternal life since karmic beings have always existed as individuals and will continue to do so, albeit in changing forms. Mahayana Buddhism places a more positive gloss on the nature of final nirvana and regards it as a state of boundless life which is not of an ontologically different order from the life we now enjoy. The crucial difference between life as we now experience it and nirvana is that the latter is not punctuated by birth and death or afflicted by the kinds of suffering mentioned at the start of this chapter.

Knowledge

The second of our three basic goods is often mentioned in Buddhist literature and the importance of knowledge in Buddhism is difficult to overstate. By 'knowledge' is meant the pure or theoretical knowledge which is known in Buddhism as *paññā* or *prajñā* (the equivalent of the Greek *sophia*) and which has as its object the truths of Buddhist doctrine. Knowledge is the result of the proper use of the intellectual capacities in reasoning, reflection, meditation, logic and so forth. Buddhism does not value all knowledge equally and would subordinate general knowledge (information) to theoretical knowledge of its teachings. Knowledge *par excellence* is therefore knowledge of Dharma. In view of the great emphasis placed on it it is reasonable to conclude that Buddhism sees knowledge as a basic good. Some, indeed, have seen Buddhism purely and simply as a way of knowledge. This is unsatisfactory in view of the abundant evidence that Buddhism recognises other human goods besides intellectual ones.

Where ethics is concerned, knowledge has both a theoretical and an applied role. The theoretical role of knowledge is to determine what is good; the applied role of knowledge is to deliberate how it should be attained. Buddhism is a cognitive ethical philosophy in that it maintains that the truth about right and wrong is objective and can be known through the proper use of the intellectual faculties. If this is so, it means that all who reason rightly will come to the same conclusions on ethical issues. We are told that all the Buddhas of the past, present and future teach the same Dharma; this means that the ethical truths of Buddhism never vary since they have been perceived to be the same by different enlightened subjects at different times. Through the use of reason individuals can ensure that the choices they make are objectively valid; this is to say that they reach the same conclusion as would an enlightened reasoner. Having said that, it must be pointed out that the goal for Buddhism is not simply to attain *knowledge* about ethics but to live a moral life in the fullest sense, and this includes having the proper motivation for ethical conduct. Thus one must both strive for knowledge about what is ethically right and also have a commitment to realising the goods which knowledge discloses. One of the most important of these is our relationship with other beings, which brings us to the third basic good – friendship.

Friendship

As noted above, in Buddhist sources knowledge is normally paired with compassion (*karuna*) rather than friendship. However, it is friendship which provides the paradigm for interpersonal relationships in Buddhism. The reason why compassion is so prominent in Mahayana Buddhism is that it is raised to the status of what might be termed a 'metaphysical' virtue. The implied compassionate subject is one of the great bodhisattvas, such as Avalokitesvara, and the assumed objects of compassion are all living creatures. Here, all are seen as appropriate recipients of compassion because from the standpoint of the enlightened all are in need of assistance. There is a danger, however, in concluding from the example of the great bodhisattvas that compassion is the paradigm virtue in Buddhism and should always play a dominant role. It must be remembered that compassion (*karuna*) in the Mahayana means two things: the first is the virtue of compassion itself, that is to say, the disposition to identify affectively with the suffering of others. The second sense of compassion is as the partner of *Prajña*, where it stands for the *totality* of moral virtues which an enlightened person possesses. Buddhas and bodhisattvas are not *just* compassionate but possess many other virtuous qualities such as generosity (*dana*), and patience (*ksanti*), which are of no less importance.

The early sources show less of a tendency to subsume the various dispositions under the heading of a single virtue, and there are several terms which distinguish the nuances within that complex of attitudes which the Mahayana denotes by 'compassion.' Four important terms are: benevolence (*metta*), compassion (*karuna*), rejoicing in the good fortune of others (*mudita*), and equanimity (*upekkha*). These four dispositions are to be cultivated and deepened in practice in daily life and through meditational exercises. Indeed, a large part of Buddhist meditational theory and practice is devoted to their development in order to counter unfriendly dispositions such as hatred.

The importance of friendship

There can be little doubt that friendship plays a central role in the
Buddhist religious life. It has been suggested that the proper way to
regard the entire human community is as a society of friends.
According to Phra Rajavaramuni, 'We might conclude that in
Buddhist social ethics everyone is a friend, meaning that everyone
should be treated as a friend.'[85] The Buddha himself said, 'My
livelihood is bound up with others,'[86] and on another occasion
stated, 'It is owing to my being a good friend to them that living
beings subject to birth are freed from birth.'[87] When asked about the
role of friendship in the religious life he rejected the view that
friendship was merely half of it insisting that it was the whole of
it.[88] The qualities of true and false friends are set out in the *Advice to
Sigala,* and four examples are given of the positive facets of friendship. There is the friend who provides help and assistance
(*upakara*), the friend who is constant (*samana-sukkha-dukkha*), the
friend who gives wise counsel (*attha-akkhayi*), and the friend who is
compassionate (*anukampaka*).[89] It is when we see others as our
friends that we come to share their joys and sorrows, to offer our
help and support, and become in all respects well disposed
towards them. To see friendship as the paradigm for moral
relations is arguably more appropriate than focusing upon a single
sentiment such as compassion. Relations with others can take
infinitely complex forms, and different attitudes and responses will
be appropriate on different occasions. When friends experience
good fortune the appropriate response is to share their happiness,
not to feel compassion for them. When they need help in a project
of some kind we respond by going to assist them with our labour
or ideas. It is only when they suffer a misfortune that compassion
for their suffering would be appropriate. The main elements of
friendship might be summed up as an attitude of benevolence and
goodwill to our fellows, a desire for peace and harmony amongst
them, and also the disposition to act justly towards them by placing
their interests on a par with our own.

Friendship as reciprocal

Friendship as a basic good has a much broader meaning than
friendship in its everyday sense. Friendship normally involves not
just the disposition towards friendship (friend*liness*) but also *being*

friends with particular individuals. Aristotle points out that friendship is normally thought of as a relationship between specific individuals involving some degree of reciprocation:

> Those who wish for the well-being of others ... are called well-disposed if the same feeling is not evoked from the other party, because goodwill, they say, is friendship only when it is reciprocated. Perhaps we should add 'and recognized'; because people are often well-disposed towards persons whom they have never seen ... but how could we call them friends when their feelings for one another are not known?[90]

In normal society no one can be friends with everyone else in this reciprocal sense. While a person can be disposed to friendly relations, it is a contingent matter whether friendships actually come about. Inevitably, there will be constraints on how far an individual can realise a particular good, such as friendship, in his or her chosen form of life. Some life-styles, such as living in a community, offer greater opportunities for friendship than others, such as living as a hermit. While it is true that the unreciprocated benevolence of the hermit cannot be called friendship in the narrow sense, Buddhism seems to suggest that some form of spiritual communion analogous to friendship may be experienced through meditation. Those who meditate on benevolence, it is thought, are reaching out to others in a way which is more than metaphorical, and there is no reason why others should not reciprocate in kind. This would still fall short of friendship strictly defined so long as the parties remained unknown to one another, but at a more advanced stage of development this limitation may disappear. According to traditional Buddhist thought, spiritual development brings with it the capacity for communion on a psychic level. The ultimate goal of Buddhism could then be conceived of as one of universal friendship in which all beings are in eternal loving communion in the strict sense of 'friendship' described above. Prior to the attainment of this goal the unenlightened would be recipients of the benevolence of the enlightened but able to reciprocate only imperfectly to the extent that their individual capacities allowed.

Goods and moral choices

A teleological theory of ethics, such as we believe Buddhism to hold, bases its judgements about what is right and wrong on its beliefs about what is good or valuable in human life as an end to be pursued. Each of the three basic goods represents legitimate ends to pursue. Thus, it is good to stay healthy in mind and body, to further knowledge, and to deepen and broaden friendship with others. Human fulfilment, or enlightenment, is nothing other than the realisation of these goods. The range of possibilities opened up through the identification of the basic goods presents a number of problems and challenges. It is at this point that we glimpse the moral dimension associated with them, for now choices must be made about what should and should not be done in their realisation. An important methodological principle follows from the recognition of certain things as basic human goods. Although the basic goods are not moral values in themselves, the very acknowledgement of them as good carries the normative implication that it would be wrong to oppose them. How might they be opposed? Since ethics concerns choices, the only way they *can* be opposed is by making choices which negate them. The most common examples of choices made against the basic goods are none other than the very acts prohibited by the precepts. The search for the principles which lie beyond the precepts has come full circle, and we can now see more clearly what it is that the precepts defend and why.

Deriving the precepts

In terms of the four-storeyed structure with which we began the discussion we find the basic goods themselves occupying the highest level. Beneath them are the methodological principles which determine how the basic goods should be pursued; that one should never directly choose against a basic good is an example of such a principle. These principles which are found at the third level may be described as 'intermediate', in so far as it is through them that the precepts are derived from the basic goods. This relationship might be represented in the form of a deductive argument which has a particular precept as its conclusion. The first premise of the argument would be that '*x* is a basic good'. The second premise would be supplied by the principle that it is always wrong to choose against a basic good, and the conclusion would be a rule that certain kinds of acts should therefore not be done. If we take the good of life as an example the steps in the argument which would lead us to derive the first precept would be: (i) life is a basic good; (ii) it is always wrong to choose against a basic good; therefore (iii) one should never choose against life. The most obvious way in which one might choose against life is by acting so as to intentionally bring about its destruction, and this is exactly how the first precept is formulated. Intentionally to deprive someone of life is to deny that life is a basic good, since the one who takes life regards it not as intrinsically valuable but as something with a relative value. In effect the one who takes life has put a price on the head of the victim by comparing it to some competing interest and finding their life to be worth less. It is this kind of computation which characterises all forms of utilitarianism.[91]

The good life for man

In terms of the broader ethical question of the right kind of life to lead, we can see that such a life will be one which does not negate any of the basic goods and is in principle open to participation in all of them. In the present context this means that a person's life should be open to participation in the goods of life, knowledge and friendship. In terms of traditional Buddhism, these goods are participated in through following the Eightfold Path. Reason requires that in the pursuit of human fulfilment none of the basic goods should be deliberately subverted. This does not mean that they must all be pursued equally, for individual circumstances and aptitudes may lead to a career or vocation in which some are more readily participated in than others. On a day-to-day basis the decision to realise one good may preclude another; time spent with friends, for example, may reduce the opportunity for study. In cases such as this, however, the neglect of one basic good comes about as an undesired side-effect of the choice to realise another.

Consequentialism

If the above account of the structure of Buddhist ethics is correct it provides further evidence that Buddhist ethics cannot be consequentialist. We saw above that the three basic goods are irreducible, which means that they cannot be boiled down, reduced, aggregated or 'netted' in any intelligible way. This is, however, precisely what all forms of consequentialism require. Consequentialism proceeds on the assumption that the consequences of an action can be reductively netted and given a value in terms of a single common denominator (for example, 'pleasure'). With respect to Buddhism such a procedure could never succeed, for what would be the utility to be maximised? Would it be knowledge, or friendship, or life? And if there is a choice between them, how is one to be weighed against the other? Should friendship be sacrificed to knowledge, or vice versa? The fundamental problem here is that it is impossible to quantify these things and trade them off against one another as if they could be related on a common scale. To do so would be like comparing chalk and cheese. None of these things can stand as 'greater' in relation to another which is 'lesser'. Because of this incommensurability between the basic goods no consequentialist reading of Buddhism could be successful, since it is the essence of any such system to compute net balances of 'good' as a preliminary to making moral choices. Since it would be senseless to attempt a comparison of incommensurables, there is a fundamental methodological obstacle in the way of characterising Buddhist ethics as 'consequentialist.'

Relativism

A further point concerns the universal nature of the basic goods. If intelligence recognises certain things as basic human goods (the paradigm intelligence here is that of an enlightened reasoner such as the Buddha) then they are good for everyone and not just for me. This means there is an obligation to promote the good of all concerned rather than just the good for ourselves. If one truly respects the basic goods as aspects of human flourishing he or she will always be disposed to co-operate with others to bring about their realisation. The basic goods must be viewed not just from our own perspective but from an impersonal standpoint such that we desire that everyone should participate in them. It might be said that we have a duty to help others participate in the basic goods and that we should be impartial as to their realisation. Immoral action fails in this respect and is always self-centred in seeking to close off or exclude the participation of others and reserve some aspect of the good to ourselves alone. One implication of this interpersonal and universal dimension of the basic goods is that just as it would be wrong for *me* to negate the basic goods then it is wrong for *anyone* to negate them. A relativistic understanding of morality as based on personal or cultural preferences is accordingly ruled out. This is confirmed by the early sources which clearly understand Buddhist morality as authoritative for all. In the *Lion's Roar of the Universal Monarch*, for example, the righteous Buddhist ruler (*cakkavatti*) is depicted as spreading Buddhist teachings in the four directions across the face of the globe. The different regents he encounters all seek instruction from him. The teaching he gives, and which they accept, is none other than the Five Precepts.[92]

We can also see from the above how the universal and impersonal features of the basic goods coincide with the doctrine of no-self. The doctrine is affirmed by the manner in which we universalise ourselves in the pursuit of the basic goods. By participation in each of them we transcend the limitations of individuality in a particular way. Through knowledge we come to know the Dharma and identify ourselves intellectually with truth. Through friendship we project our affections into the universe of living beings and enter into communion with them. Ultimately, in final nirvana we escape the limitations of individual existence and experience the goods of life, knowledge and friendship in a boundless way that it is difficult to comprehend at the present time.

VI   A TEST CASE

The above account of the principles underlying Buddhist ethics is rather compressed, and leaves open many questions which need to be addressed further. In particular, we may wish to propose a more comprehensive list of basic goods by extending the list of three mentioned above. There is also scope for extending the list of methodological principles of practical reason of the kind which supplied the second premise to our example above ('never choose directly against a basic good'). We have limited the discussion here to what seemed most relevant to the matter in hand, namely the attempt to elucidate the underlying values of Buddhist ethics in order to allow us to move beyond the basic formulation of the precepts (particularly the first precept) to a level from which we can address biomedical issues of a complex nature.

Scriptural justification

When discussing methodology earlier it was suggested that any view which purports to be 'Buddhist' must offer scriptural justification for its opinion. It is therefore time to put the theory set out above to the test and find out how well it stands up to the evidence of our sources. Since it is extremely unlikely that there exists

anywhere in Buddhist literature a theoretical account of Buddhist
ethical principles to which we could compare it, we are forced to test
the theory by more indirect means. Two criteria by which a theory
might be assessed are its explanatory power and the absence of
counter-examples to its predictions. Counter-examples to the theory
proposed would take the form of textual evidence to the effect that
Buddhism does not value life, knowledge and friendship. Since
Buddhist literature everywhere promotes these values, however, it
is unlikely that any such evidence to the contrary will be found.

A counter-example: suicide

It might be suggested that a counter-example to the claim that life is
a basic good is provided by the instances of religious suicide
described in Buddhist sources. Wiltshire draws attention to three
cases in the Pali canon as 'instances of suicide which, if not
condoned, are certainly exonerated',[93] and Lamotte cites examples
of meritorious self-immolation from later Buddhist literature.[94]
Suicide in Buddhism (and other Indian religions) was also
commented on by Durkheim in his classic early study.[95] After
surveying the evidence Lamotte concludes that the examples of
suicide in Buddhism can be explained by reference to one of three
motives.

> In brief, then, if suicide was practised widely in Buddhist circles,
> this was due to three reasons. [1] In the Hinayana the Noble Ones
> – whether Buddhas, Pratyekabuddhas or Arhats – once their
> work was done, met death voluntarily in order to enter Nirvana
> as soon as possible. [2] In the Mahayana, the Bodhisattvas offered
> up their bodies and lives for the welfare of beings or [3] in order
> to pay homage to the Buddhas.[96]

Do the deaths which occurred in these circumstances show that
Buddhism does not respect life as a basic good? It would not appear
so since it is by no means clear that any of them involve a choice
against life *tout court*. The deaths occur in exceptional circumstances
and all involve an extraordinary degree of saintliness or religious
piety. They have little in common with the circumstances in which
most suicides take place. The standard Buddhist attitude towards
suicide is that it is a futile, misguided act motivated by the desire for
annihilation (*vibhava-tanha*).

What we see in the examples above is not suicide as normally understood (and condemned), but a special category of voluntary death motivated by religious zeal.[97] The deaths are more analogous to martyrdom than suicide, and involve not so much the denial of life as the affirmation of a higher spiritual ideal. In the first type of case cited by Lamotte, the moral choice involved is best described not as a choice against life but a choice in favour of nirvana. The cyclic existence of an enlightened person is at an end, so the option of choosing either life or death in the sense which this choice is normally available, no longer arises. The choice instead is whether to enter nirvana now or later. We have described nirvana as the fullest and most perfect form of life, so the affirmation of nirvana cannot be a choice against life. It must be added that the textual evidence that suicide in such cases is either condoned or exonerated is less than compelling.[98]

The second motivation described by Lamotte is altruism. Although the circumstances in which bodhisattvas sacrifice themselves are of an extraordinary kind, the moral issue is relatively unproblematic. Bodhisattvas who sacrifice themselves are not choosing against life but displaying a readiness to lay down their lives in the service of their fellow man. They do not seek death for its own sake, but accept that death may come, so to speak, in the course of their duty. Their actions can be compared with those of a soldier who throws his body on a grenade to protect the lives of his comrades. The soldier is not seeking to end his life, but willingly accepts death as the price of his action. If he found the pin had been left in the grenade, he certainly would not remove it himself.

The third type of motivation described by Lamotte refers to cases such as that of the bodhisattva Priyadarsana in the Lotus Sutra. Lamotte described how, as an act of homage, Priyadarsana:

... spent twelve years in ceaselessly and constantly partaking of inflammable substances. At the end of those twelve years, Priyadarsana, having clothed his body in heavenly garments and sprinkled it with scented oils, made his benedictory aspiration and then burnt his body, in order to honour the Buddha Candrasurya and the discourse of the 'Lotus of the Good Doctrine'.[99]

The text states that the ensuing pyrotechnics lasted for twelve hundred years. I do not propose to offer a justification for this last category of voluntary death since fictitious examples of this kind typify a genre of devotional literature which does not have normative ethics as its main concern. We conclude that all three of the above categories are exceptional in one way or another, and that it would be unwise to treat them as having normative relevance as far as ethics is concerned. Assuming, then, that the examples of suicide discussed above do not disturb our claim that Buddhism regards life as a basic good, the theory may be said to have passed the first test (no convincing counter-examples).

Explanatory power

We now move on to an assessment of the theory in terms of its explanatory power.[100] The theory offers a justification for the Buddhist precepts which is more comprehensive than anything found in the sources. Perhaps the boldest claim made by the theory is that what is morally right is to be determined by reference to the basic goods rather than to factors such as motivation or consequences, either together or by themselves. The acid test of the theory would therefore be a case which was inexplicable by reference to motive or consequences alone. Such a case would have to meet three conditions: (i) there was a good motive; (ii) the consequences aimed for came about, but (iii) the act was nonetheless declared to be wrong. If we could find an example of such a case in Buddhist literature, preferably canonical and involving the Buddha himself, it would eliminate the two main candidates which have presented themselves so far as grounds for moral validation. If we could show further that our own account could explain why the action was immoral where the others failed to, we would have strong grounds for believing it to be true.

The case

We are fortunate that we find a case fulfilling all of these requirements in the Monastic Rule. The case concerns a disciplinary matter upon which the Buddha himself pronounces judgement. In fact it is the very first case cited under the third *parajika* or offence of 'Defeat', namely 'depriving a human being of life'. The case fulfils the conditions we require and turns upon the doing of a wrong deed from a good motive. It is very short and is cited in full below:

> At that time a certain monk was ill. Out of compassion the other monks spoke favourably to him of death. The monk died. The

others were remorseful and wondered: 'Have we committed the offence of Defeat?'[101] These monks reported the matter to the Lord. 'Monks', he said, 'You are guilty of the offence of Defeat'.

From these few terse lines we could without too much difficulty reconstruct in our minds the all too human scene, one re-enacted in hospitals and hospices every day. A monk was ill, apparently gravely so, probably in great pain and with little prospect of recovery. His fellows stood by distraught yet unable to do anything to relieve his suffering. We need not rely entirely on our own imaginations since in his commentary on the case Buddhaghosa fills in the details for us:

> Turning now to the accounts of the disciplinary cases, in the first case 'out of compassion' means that those monks, seeing the great pain the monk was in from the illness felt compassion and said to him: 'You are a virtuous man and have performed good deeds, why should you be afraid of dying? Indeed, heaven is assured for a virtuous man at the very instant of death.' Thus they made death their aim and, although ignorant of the state of being one who makes death his aim,[102] praised death. That monk, as a result of them praising death, ceased to take food and shortly after died, and because of this they committed an offence … A wise monk, therefore, should not now speak in praise of death to a sick monk for if he, hearing that praise, dies through some means like ceasing to take food when even one second of life remains, then he has been killed by that person.[103]

The problem

The problem which this case presents is to explain what, precisely, the monks had done wrong to merit being pronounced guilty of the most serious category of monastic offence for which the penalty is lifelong excommunication. Their motivation is not in question since we are explicitly told that they acted out of compassion (*karuññena*). In normal circumstances this motive is highly commendable, but here it does not exonerate the monks of guilt. If rightness is determined by motivation, however, all they would seem to be guilty of is compassion for the suffering of a fellow monk. It seems, then, that the judgement in this case is inexplicable by reference to motive alone.

Similar problems are encountered with respect to consequences. The desired consequences came about and the monk was released from his suffering. This was exactly the compassionate outcome the monks desired, yet they were still judged to be guilty of an offence and excommunicated. It might be thought that the act was wrong because it generated evil karmic consequences which would take effect at some future time. This simply begs the question, however, for we should then be forced to ask why an act of compassion should have bad karmic consequences. Judging by the decision, then, neither the motive nor the consequences seem to have figured in the Buddha's assessment of the morality of what was done. The only alternative explanation for the Buddha's decision would seem to be that the act was in breach of the precepts. However, this explanation does not go to the moral status of the act in any meaningful way. It tells us only that the act is wrong because there is a precept against it, and such an explanation would reduce Buddhist ethics to little more than rule-worship.

A solution

Neither of the above explanations seems adequate to explain the judgement in the light of the facts of the case. If we analyse this decision in terms of our theory of the basic goods, however, a more satisfactory explanation may be forthcoming. On this account the act was wrong because the guilty monks intentionally chose against a basic good, namely life. The essence of their wrongdoing, as Buddhaghosa points out, was that *they made death their aim.* However benevolent their motive, nothing could legitimate this direct choice against a basic good.

To understand the decision in this case it may be helpful to draw a distinction, as the law does, between motive and intent. Motive may be defined as 'That which moves or prompts a person to a particular course of action, or is seen by him as the ultimate purpose or end he seeks to achieve by that action.'[104] It can be distinguished from intent, which is the more immediate aim to which an action is directed. This distinction is made clear in the following legal opinion:

It is, I think, important that we should know what we are talking about when we use the word 'intention.' It is obvious that it is to be differentiated from *motive*. If I kill you for your money, my intention is to kill you but my motive is to lay my hands on your money. So also, if I kill you from the motive of compassion (so-called mercy killing) I nevertheless intend to kill you and the crime is one of murder.[105]

Applying this distinction to the case in point, the *motive* of the monks was the compassionate one of ending the suffering of the sick monk, while their *intent* was that he should die. While motive is of great importance in Buddhist ethics it does not by itself guarantee moral rightness. If it did, it would be impossible to do wrong from a good motive. We see here that the Buddha felt this was only too possible, and a little reflection on history would throw up many further examples. In the case above, death was chosen as the means to bring about the end of suffering. In the light of the decision we must conclude that a course of action cannot be right if it involves intentionally turning against a basic good either as an end or as a means to an end. The implication of this case for our present concerns could be summed up in the following principle: *Karmic life must never be destroyed intentionally regardless of the quality of motivation behind the act or the good consequences which may be thought to flow from it.* This both restates and amplifies the first precept and will form the basis of our approach to the bioethical issues discussed in the following chapters.

The case in point is doubly relevant since we see that the actions of the monks were in fact inspired by another basic good, that of friendship. This manifested itself in the form of compassion for the sick monk. Nevertheless, the fact that the monks willed his death amounted to a deliberate choice against the basic good of life, and it was for this reason they were judged to have done wrong. What the monks in this case had failed to do was to respect all of the basic goods equally. Instead, by subordinating life to friendship, they chose to realise one of the basic goods by denying another. Since the basic goods are of equal importance to human flourishing, however, they cannot be traded off against one another in this fashion.

From the above discussion we can understand a little more clearly how within the Buddhist moral landscape the contours of motives, intentions, actions, and consequences come together. On the one hand, we cannot say that for Buddhism only motive counts, since the moral status of an action is not determined exclusively by motivation. For a course of action to be right it must proceed from a good motive and not involve as part of it an intention which is contrary to the basic goods. A good motive is thus a necessary but not sufficient condition for a moral act. An intention itself is right when it flows from a good motive and is in harmony with the basic goods. In the final analysis, the basic goods, motives, intentions and consequences all intersect, and from a commitment to and proper orientation towards the basic goods all manner of auspicious consequences flow.

2

At the Beginning of Life

Introduction

Understanding the origin of human life, as well as being a fascinating scientific enquiry, is also a profoundly important exercise from the point of view of ethics. The view we hold as to when individual life begins will influence the moral judgements we make when dealing with the complex ethical issues surrounding life in its early stages. As we shall see in Chapter 3, it will also have a bearing on how we define the end of life. We begin this chapter with a review of the account of the beginning of life found in traditional sources after which we will pause to reflect on its implications and its compatibility with current scientific knowledge. This will lead in Section III to a discussion of some possible objections to the Buddhist account of when life begins. After addressing these questions we will turn in Sections IV, V and VI to the substantive issues of abortion, embryo research and fertility control.

I WHEN DOES LIFE BEGIN?

A distinctive feature of Buddhist thought is that it does not postulate an initial starting point to the series of lives lived by an individual. Instead, it regards the cyclic course of human existence as potentially eternal: it had no beginning and there is no certainty it will ever have an end. What takes place at conception[1] is the *rebirth* of a previously existent individual. All conception is thus re-conception. The belief that each individual exists prior to conception provides a distinctive perspective on the question of when life begins. We must therefore begin our account of when life begins with a consideration of the state *prior* to conception. Following this we turn to the details of conception and gestation found in the early sources.

The intermediate state

Some schools of Buddhism hold that rebirth follows instantaneously upon death, while others believe there is an intermediate state which functions as a buffer between lives. The Theravada holds the former view and sees death and rebirth as a seamless continuum. It pictures the transition between the two as 'like a man who crosses a river by hanging onto a rope tied to a tree on the near bank.'[2] The Tibetan tradition, on the other hand, believes there is an intermediate state between one life and the next which it designates by the name *bardo* ('interval'). An account of the intermediate state is given in the well-known *Tibetan Book of the Dead*, and much new information on these traditional teachings has come to light in recent years with the diaspora of Tibetan lamas to the West.[3] The Bardo teachings analyse the principal human psychological experiences in both the embodied and disembodied states into a series of six *bardos*. Three of the six *bardos* are experienced in the sequence of death and rebirth: first the process of death itself, second the experience of the intermediate state, and third the search for rebirth. It is with the last of these that our account begins.

In the Tibetan sources it is said that when the time of rebirth draws near, the lights of the six possible realms of rebirth (such as heaven, hell, human, or animal) 'shine forth' to the being about to be reborn. The light of the human realm appears as blue in colour, and one who is to be reborn as a human being will be attracted by this light and gravitate towards it. This attraction is explained as a kind of libido: so long as there remains a residue of desire, the spirit of the departed person will be drawn back to the material world. Two versions of events are found which explain the impetus to rebirth in slightly different ways. In the first, which is less common, rebirth is depicted as a flight from the bewildering experience of the intermediate state, and the womb appears as a warm, safe haven. The second version explains the motivation for rebirth as desire for sensual experience. It is said that the intermediate being 'sees' and is attracted by the erotic activity of the couple who are to be its parents. In an interesting anticipation of psychoanalysis it is held that a person who is to be reborn as a male will be attracted to the mother and feel aversion for the father, and vice versa for one who is to be reborn as a female. The sudden arising of these strong feelings of attraction and aversion marks the final commitment to rebirth, since the moment in which they arise is the moment at which the intermediate state terminates and the being enters the womb.[4]

There are divergent opinions as to the actual means of ingress to the womb. In different texts the spirit is said to enter by different means, according to the spiritual status of the person reborn: through the top of the head, in through the mouth of the male and out through the penis, or directly into the vagina of the female. It is also pointed out quite logically that since an intermediate being is unobstructible it does not depend on any single means of ingress. The disembodied consciousness in the *bardo* is likened to molten iron, and its embodiment at conception to the cooling and hardening of molten iron into a definite form.[5] The actual experience of entering the womb varies according to one's karmic status:

> A person of little merit hears clamorous noises and has a sense of entering into a marsh, dark forest or the like; whereas one accustomed to good deeds hears peaceful and pleasant sounds and has a sense of going inside a nice house, etc.[6]

Bardo symbolism

The descriptions of *bardo* experiences can be regarded as either veridical or symbolic. Upon reflection it can be seen that the more colourful accounts of the rebirth experience can be read as parables which illustrate doctrine. Buddhist teachings hold that greed, hatred and delusion are the roots of all evil. These are the things which fuel the round of *samsara* or cyclic rebirth. The account of the experience of the intermediate being shows in graphic detail the part played by these defilements. To give them a precise role in the actual mechanism of rebirth, as when the *deluded* intermediate being experiences *desire* for the material world and feels *aversion* to the intermediate state; or by speaking of *desire* for one parent and *aversion* to the other, is to make explicit the connection between the three 'roots of evil' and rebirth, thereby validating doctrine. The vision of the parents in the act of procreation is also consistent with the Buddhist distrust of sex, which it regards as one of the most powerful sources of attachment. It is the physical act of sex which brings about rebirth with all its attendant ills, and here it is the vision of this *particular* act of intercourse which marks the entry to a new cycle of life for the individual. The accounts of the experience of the intermediate being can thus be seen as providing vivid symbolic confirmation of the truth of Buddhist teachings.

The Buddha himself does not discuss the intermediate state, except to say that he could perceive through his clairvoyant power the course of beings dying in one life and arising in the next. He compared this to watching a man leave one house and enter another. His reticence on the matter is quite consistent with his overall pragmatism and his primary concern with *this* life, rather than those which had been or were to come. It is difficult to tell, therefore, to what extent the accounts of the experiences of the intermediate state are imaginative reconstructions to illustrate doctrine, or are based upon memories of personal experiences. According to Buddhist psychology the memory-traces of all experiences, including those of previous lives, are retained at a deep level of the mind, and there is no reason in principle why advanced yogins should not be able to recall and describe in detail an experience such as that of the intermediate state if it occurred. Whatever the exact nature of this interlife transition, its terminus marks the beginning of a new individual life.

Conception

Although the basic Buddhist position on how and when individual life begins was formulated over two thousand years ago, the conclusions reached are in some respects remarkably modern. Buddhism has always seen conception as an event marking the start of an orderly process of development up to birth and beyond, through childhood into maturity. In adopting this view Buddhist thinking was very much ahead of the West, and its views on embryological development, although not accurate in all respects, are broadly in line with the discoveries of modern science. Western thinking on the matter followed Aristotle for almost two thousand years, and its understanding of embryology was hampered by his theory of progressive animation through the sequence of vegetative, animal, and rational souls.[7]

The Buddha explained conception as a natural process which occurs when a specific set of conditions is fulfilled. However, he did not explain the process solely in biological terms. Although the necessary biological conditions must be met he also made reference to what Soni has termed 'ultrabiological' factors.[8] As with Aristotle, Indian thought understood the substance of the conceptus as constituted by the mingling of sperm and menstrual blood. Unlike Aristotle, however, it postulated a spiritual counterpart as necessary for the generation of human life. Jolly has provided the following statement which summarises the opinion of the classical Hindu medical treatises as to how conception occurs, and which seems to express a pan-Indian view:

> In the union of husband and wife ... the sperm comes out ... arrives in the uterus and is united with the menstrual blood. Thus the foetus is created when the spirit (*jiva, cetanadhatu*), quick as wind and impelled by his deed in an earlier birth (*karman*), enters it as the sixth element. If the sperm preponderates a male child is born, if the menstrual blood prevails a female child is generated ...[9]

The three conditions

The Buddhist understanding of conception is similar to the account given above, and in an early canonical text the Buddha explains the conditions under which it occurs. The passage sets out three conditions which are required if an intermediate being is to enter the womb and embark upon a new human life. This passage and other early texts speak of the 'descent' or 'entry' (*avakkanti*) of the intermediate being into the womb.[10]

> Monks, it is on the conjunction of three things that there occurs the descent of an intermediate being[11] into the womb. If the parents come together in union, but it is not the mother's proper season, and the intermediate being is not present, then there will be no conception. If the parents come together in union and it is the mother's proper season but [still] the intermediate being is not present [again] there will be no conception. But when the parents come together in union, it is the mother's proper season and the intermediate being is present, then on the conjunction of these three things the descent of an intermediate being will take place. Then, monks, the mother for nine or ten months carries the fetus in her womb with great concern for her heavy burden.[12]

Thus the three conditions to be fulfilled are: (i) intercourse must take place; (ii) it must be the woman's fertile period; (iii) there must be an intermediate being available to be reborn. It is interesting to note that the Buddha's statement on the 'conjunction of three' above is repeated on another occasion in the Pali canon by Brahmins,[13] which suggests it was part of common Indian lore. The *Caraka Samhita*, a classical Hindu medical treatise dating from early in the common era, gives an account similar to that above:

> Conception occurs when intercourse takes place in due season between a man of unimpaired semen and a woman whose generative organ, (menstrual) blood and womb are unvitiated – when, in fact, in the event of intercourse thus described, the individual soul (*jiva*) descends into the union of semen and (menstrual) blood in the womb in keeping with the (*karmically* produced) psychic disposition (of the embryonic matter).[14]

We see here again reference to the same three conditions, namely: (i) intercourse must take place; (ii) it must be in 'due season', i.e. at the appropriate phase of the menstrual cycle; and (iii) the spirit of a being seeking rebirth must be at hand. It could be suggested that a fourth condition has now been added, namely that there be no defect in the reproductive capacity of either parent. Most probably, however, this should be taken as implicit in the requirements laid down by the Buddha. This would seem to be confirmed by a later Tibetan source which makes this condition explicit:

> First, the causes of formation [of the foetus] in the womb are: non-defective sperm and blood of the father and mother, the consciousness [of the being who is about to enter the mother's womb] that is impelled by karma ... mental distortions, and then the collection of the five elements.[15]

We find further agreement with reference to the details of the female reproductive cycle alluded to in the Buddha's second condition. In his commentary on the Buddha's statement regarding the three conditions, Buddhaghosa gives us an insight into early embryological thought:

> *It is the mother's fertile period.* This is said with reference to the fertile time. In women, in the place where a child is reborn in the womb, a large pustule of blood collects then breaks and issues forth. The site is then pure. When the site is pure and the parents come together once in union then the site becomes a fertile field for seven days and then the child begins to grow acquiring bodily attributes such as hands and hair, etc.[16]

According to the material collated by Jolly from the major Hindu medical treatises the period suitable for conception is the twelve nights after the beginning of the menses. The first three or four must be excluded since semen which arrived during the bleeding in the uterus would be swept away like an object thrown into a flowing stream. This would leave a fertile period of either seven or eight days, closely in line with Buddhaghosa's estimate above. The whole of the fertile period from the onset of menstruation is known as the *rtu* or 'season'. At the end of the *rtu* 'the uterus does not allow the sperm to penetrate, just as the lotus closes at the end of the day.'[17]

Gestation

The Buddha divides the stages of childbearing into four: the fertile period, pregnancy, birth, and nursing.[18] The descent of the intermediate being into the womb occurs during the first of these and marks the point of origin of a new individual life. The gender of the individual was determined at conception[19] and from this point onwards the material and spiritual components which constitute the new individual evolve together. These two components, body and spirit, are sometimes compared to two sheaves of reeds in a field which lean on one another for support.[20] Other images, suggestive of a dualistic understanding of the relationship, compare the spirit to a jewel wrapped in cloth[21] and to a house which the spirit enters and inhabits:

The material body, householder, is the home of consciousness (*viññana*). So consciousness, which is tied by greed to the material body, is called the dweller in the home.[22]

From a doctrinal point of view conception is an event of great significance, since it marks the beginning of another cycle of life with all its attendant ills. It is the event *par excellence* which Buddhist teachings set out to prevent, and the Buddha gives a causal analysis in philosophical and psychological terms which explains how beings come to be repeatedly entrapped in the cycle of re-conception (*samsara*). Below, he describes to his disciple Ananda how the 'descent' of consciousness into the womb triggers and sustains the development of life:

I have said that in dependence on consciousness (*viññana*) there is mind and body, and that, Ananda, is to be understood in this way:

were consciousness not to descend into the mother's womb, would mind and body become constituted there? They would not, Lord. Were consciousness, having descended into the mother's womb, to become extinguished, would mind and body come to birth in that state of being? They would not, Lord. Were consciousness to be extirpated from one who is young, either a baby boy or girl, would mind and body attain to growth, development and expansion? They would not, Lord. Therefore, Ananda, consciousness indeed is the cause, the ground, the genesis, and the condition of mind and body.[23]

Embryonic development

Once consciousness has 'descended' into the womb and conception has occurred, the embryo develops through a set number of stages. In *The Path of Purification*, Buddhaghosa lists four stages of the early embryo during the first month after conception. The first stage is the *kalala*, in which the tiny embryo is described as 'clear and translucent', and is likened to 'a drop of purest oil on the end of a hair'.[24] The following three stages are the *abudda*, the *pesi* and the *ghana*, terms which connote increasing density and solidity. Sometimes a fifth stage, *pasakha*, is mentioned.[25] After the first month, the phases of pregnancy are enumerated by reference to the number of months from conception.[26] It was noted that throughout its time in the womb the child does not breathe.[27]

Further details of embryonic development are found in an authoritative seventeenth-century Tibetan treatise which quotes several earlier sources.[28] This text confirms the picture already sketched, suggesting that the early views remained influential and underwent little, if any, modification. It relates how following intercourse the 'drops of semen and blood ... are mixed in the mother's womb', and the consciousness of the intermediate being enters into this mixture. 'Initially', we are informed, 'the oval-shaped foetus is covered on the outside by something like the cream on top of boiled milk; but inside it is very runny.'[29] It was thought that 'The place in the semen and blood where the consciousness initially enters becomes the heart',[30] and that in this initial phase of development 'The top and bottom [of the body at this point] are thin, and the middle is bulbous like the shape of a fish'.[31] The text goes on to describe the course of development within the next twenty-eight days:

When the oval-shaped foetus has passed seven days ... [it] becomes viscous both outside and inside, like yogurt, but has not become flesh. When another seven days pass ... the foetus becomes fleshy but cannot withstand pressure. After another seven days it hardens ... [so that] the flesh is now hard and can bear pressure. When this, in turn, has passed seven days ... the foetus develops legs and arms, in the sense that five protuberances – signs of the two thighs, two shoulders and head – stand out clearly.[32]

The sources cited by this text are in broad agreement that the length of a normal pregnancy is thirty-eight weeks: one source places it at two hundred and sixty-eight days and another at two hundred and seventy days.[33] The traditional understanding of conception and embryology may thus be summarised as follows:

1. In between the woman's monthly periods one or more pustules of blood collect in the place where the embryo will be conceived and grow (the womb)
2. In the normal menstrual cycle these break and flow forth causing a monthly period
3. When the period ceases a residue of blood remains and the site of conception is fertile for a period of between three to ten days
4. When intercourse takes place the semen mingles with the menstrual blood
5. If an intermediate being is available it 'descends' into the union of semen and blood – this is 'conception'
6. The conceptus in this early stage is like a tiny drop of fluid and is known as the *kalala*
7. The embryo develops through three further stages within the first month
8. Its development continues between the second and the tenth month coming to full term after about thirty-eight weeks.

II REFLECTION ON THE THREE CONDITIONS

We now have a sufficient understanding of conception and embryology in Buddhism to allow us to proceed to a critical examination of these beliefs from a philosophical and scientific perspective. We can see that in broad outline Buddhist ideas complement modern thinking while differing in matters of detail. This is hardly surprising in view of the great antiquity of the early sources and the primitive resources available to physicians at the time. If we are to apply the principles of Buddhist ethics to contemporary biomedical problems, however, we must bring the ancient sources up to date and ask ourselves how they would express themselves with the benefit of modern scientific knowledge. The most important issue here concerns the point at which human life was thought to commence. From the traditional accounts we have some reason for locating this close to the time of intercourse, but we may now enquire whether further precision is possible.

The sequence of the three conditions

Let us begin by asking how the Buddha would have modified his statement about 'the conjunction of three' if he were addressing a modern audience. The Buddha states above that three conditions are required for 'the descent of the intermediate being'. Each of these is a necessary condition: none of them is sufficient alone nor are any two of them. Only when all three are fulfilled do we have a set of conditions which is both necessary and sufficient for conception to take place. Unfortunately, the Buddha does not specify clearly the chronological sequence of these events in connection with the fulfilment of the three conditions and the descent of the intermediate being. We can see that the descent of the intermediate being cannot take place until all three conditions are fulfilled, but what is not entirely clear is whether the descent is triggered immediately the conditions are met. In other words, is the descent of the intermediate being *simultaneous* with the fulfilment of the three conditions, or can it occur subsequent to it? The Buddha does not *say* there is any delay, but we cannot be entirely satisfied with an argument from silence. Although rather forced, it would be possible to construe the Buddha's statement as meaning that the descent of the intermediate being could occur later. This would mean that the intermediate being was available and committed to that particular rebirth without as yet having entered the womb. On this understanding the ovum could be fertilised *before* the intermediate being descended. This interpretation would open the way to the suggestion that the descent of the intermediate being occurs around the time of implantation, a process which begins about six or seven days after fertilisation.

Can we glean any further help from the passage itself? The text speaks of a 'conjunction' (*sannipata*) of the three conditions. Etymologically, this connotes a 'falling together' (*nipata*) or perhaps 'falling into place'. The prefix (*san-*) suggests a combination or juxtaposition of several elements. The word 'conjunction' is in the ablative case which implies that is 'due to' or 'because of' this conjunction that the subsequent event occurs. Buddhaghosa glosses 'conjunction' as 'collocation' (*samodhana*) or 'conglomeration' (*pindabhava*).[34] Interestingly, the word 'conjunction' (*sannipata*) occurs twice in the passage: the second occurrence describes the physical action of the parents 'coming together' in sexual intercourse. These factors suggest that the three conditions were thought of as converging or interlocking, and thereby collectively enabling or triggering the descent of the intermediate being simultaneously with their fulfilment. The image which comes to mind is that of a combination lock: only when the three combinations have been entered correctly can the door be opened.

The imagery

The passages cited from both Hindu and Buddhist sources provide details of what was thought to happen during or soon after intercourse. The sperm was thought to enter the womb where it mingled with the residue of blood from the menstrual period. When the semen and blood mingle the intermediate being fuses with this biological matrix which will constitute its physical body from that point on. This initial mingling seems to be the decisive biological factor in the generation of an embryo because it is the event to which the sources make repeated reference as the culmination of intercourse.

The imagery used suggests that the descent of the intermediate being was thought to be simultaneous with intercourse. This is shown by the intermediate being finding itself attracted to rebirth through witnessing the erotic activity of its future parents. It is represented as becoming emotionally involved in the proceedings and descending while intercourse is actually taking place. Clearly, all the excitement is long over by the time of implantation, and it seems unlikely that the intermediate being would be irresistibly attracted to a process of cell division or some other biological process.

Evidence from a Tibetan source seems to confirm the supposition that conception occurs at or soon after intercourse. The text compares intercourse to the rubbing together of two pieces of wood, and conception to the generation of fire.[35] Fire cannot be produced when the two sticks are apart, which suggests that conception was thought to occur during intercourse. The same text provides another helpful image using the generation of fire to explain how conception occurs:

> The mother's blood may be likened to a flint, the father's sperm to the iron, the consciousness that enters the mixture to a piece of bark and the embryo to the fire.[36]

Conception is here likened to the production of fire using flint, iron and bark. Perhaps a modern equivalent of this would be a gas cigarette lighter, in which flint, grinding wheel and fuel interact to produce a flame. Intercourse is like the striking together of flint and iron. While this process will create a spark it will not, in the absence of combustible material, produce fire. Only when the gas, representing the spirit of a person seeking rebirth, is present will the conditions needed to create fire be fulfilled. This image lends strong support to the view that conception was thought to occur simultaneously with intercourse, since fire cannot be produced unless all three elements interact at one and the same time. It also suggests that the fusion of spirit and matter is instantaneous rather than a process which takes some time to complete. Just as fire has either been produced or it has not, so conception has either taken place or it has not.

The first condition

We may briefly review the three conditions in turn in the light of modern knowledge, beginning with the first, namely intercourse. We now know, of course, that life is not produced by the mingling of semen and blood. Instead, the sperm comes into contact with an ovum in the oviduct or Fallopian tube. Once fertilised, the ovum travels along the Fallopian tube to the uterus where it implants. Cell division within the embryo continues during its journey to the uterus which takes about five days, at the end of which the original single cell will have multiplied to over one hundred. The embryo then touches down on the inner lining of the uterus about one week after fertilisation. In the light of this modern knowledge, when would our ancient sources say that conception takes place and new individual life begins? In scientific terms there would seem to be only two realistic candidates for this: fertilisation and implantation.

Implantation

Let us take the time of implantation first, recalling that this would only be a possibility if the evidence above concerning the timing of the descent of the intermediate being was thought unpersuasive. If we took implantation as the beginning of a new life it would mean that the new individual being came into existence at the earliest a week or at the latest about fourteen days after fertilisation. It must be said there is little to support this view other than the description in our sources of the sperm mingling with the menstrual blood in the womb, a description which might loosely correspond to implantation. One drawback with this view, from a scientific standpoint, is that implantation is a process rather than an event, and it takes about eight or nine days for the embryo to complete the process of burrowing into the lining of the uterus. Would the new life commence (i.e. would the intermediate being descend) at the beginning of this process, at the end of it, or at some point in the course of it? There seems no real reason to prefer any of these points over another, which means that the choice of any one of them would be arbitrary. However, this need not be a fatal objection, and it could be held that the new life began at *some* point during this phase, and that this point need not be the same in every case. Implantation would therefore represent a 'window of opportunity' for a new life to come into being. If it did not come into being, we must assume that the newly fertilised ovum would simply fail to develop and in due course be rejected and lost.

A further drawback with locating the descent of the intermediate being at implantation is the dualistic implication that an individual's biological nature can antedate its existence as a composite being. Here one part (the biological) would have a longer history than the other (the spiritual) in the course of the same life. However, traditional Buddhist teachings on the interdependence of the material and spiritual aspects of human nature suggest that the two arise simultaneously rather than one after the other.

Fertilisation

Are these problems avoided if we turn our attention to fertilisation as a candidate for the point at which new life begins? To place the beginning of life at fertilisation we would need to reject the image of the sperm mingling with the menstrual blood as the defining moment in the origin of new life. This does not present a serious problem as we can reinterpret this aspect of the traditional account quite easily as a reference to the meeting of the respective male and female genetic contributions in the form of sperm and ovum, rather than sperm and menstrual blood. Our sources knew nothing of ovulation, so it is quite clear that they could not have represented the process correctly. We must accept that the view of the menstrual blood as the mother's contribution was logical in the light of medical knowledge at the time.

A second argument in favour of this view is that fertilisation is much closer to intercourse than is implantation. If we have read it correctly, the Buddha's statement on the 'conjunction of three' implies that conception and intercourse are simultaneous, or nearly so. Science has shown that the two are not exactly simultaneous, and that fertilisation normally takes place from within five minutes to an hour after intercourse.[37] Again, we must allow for the ignorance of our sources about this aspect of the process. They reasoned that conception would occur as soon as semen mingled with the residue of menstrual blood. In terms of this model of conception there was no reason to envisage any delay between intercourse and fertilisation. The discovery of ovulation, however, introduces a complication which they could not have been aware of. This means that the timetable for fertilisation has to be put back slightly. In view of the importance attached to the act of sexual intercourse in the early accounts, however, an explanation which located fertilisation as close to this event as possible is to be preferred to one which places it at a more remote time.

The sources are describing, using the concepts available to them, the origin of new human life. This was understood conceptually as the point from which all subsequent development proceeds, and before which no material basis for individual life exists. Translating these requirement into modern terms we would have every reason to locate the descent of the intermediate being at fertilisation. Before this time there is no genetic individual, and after it one has come into being. All subsequent developments in the history of the individual in the present life, including implantation, can be traced back to this point but not beyond it. It is difficult to wish for a clearer point of origin, and fertilisation seems by far the most likely candidate for the point at which new life begins.

The second condition

This condition requires that the woman should be in her fertile period. It presents little difficulty, and is in accordance with modern information. Our sources place the onset of the fertile period slightly earlier than it is now known to be, and in fact it lies closer to the middle of the cycle than the beginning. However, this does not affect the basic requirement of the second condition that the woman should be in the appropriate phase of her menstrual cycle when intercourse takes place.

The third condition

It will be recalled that the third condition specifies the presence of an intermediate being. This condition is significant both in itself and in the role it plays with respect to the other two. By placing this condition third, the Buddha's statement opens the way to an interesting possibility, and to understand the significance of this we must recall our discussion of the concept of a moral 'person' in Chapter 1. Philosophers who allow rights only to 'persons' commonly argue that 'personhood' cannot be imputed in the absence of a brain and central nervous system. Since these do not appear until well after fertilisation it is suggested that the conceptus up to that point is best described as 'human biological material' or as a 'potential person'. Although we have argued that Buddhism would reject the notion of 'personhood' as having any ethical validity, there would seem to be one set of conditions under which it would agree with the view that there can be such a thing as an embryo which lacks full moral status.

One conclusion which seems to follow from the Buddha's statement is that there can exist fertilised ova which lack the essential ingredient which would distinguish them as moral beings. To see how this position arises all we need do is imagine that the first two conditions laid down by the Buddha are fulfilled without the third. This is, in fact, an eventuality which is specifically mentioned in the text, for the Buddha refers to the possibility of the first two conditions being fulfilled in the absence of the third. If the first two conditions can be fulfilled without the third, as apparently they can, it means that all the necessary biological conditions for new life as the Buddha understood them (conditions one and two) can be present in the absence of the consciousness of an intermediate being (condition three). If we translate this into modern terms it means that fertilised ova can exist which have not been animated by the consciousness of an intermediate being. This means that for Buddhism there is no reason to believe that *all* fertilised ova are new human individuals. To put the matter in terms more familiar to the debate which has been conducted on this matter within Christianity, it would mean that there is no scriptural requirement for a belief in the immediate animation of fertilised embryos.

Fertilisation and animation: a review

But perhaps our translation of the Buddha's statement into the terms of modern biology is inaccurate, and since this is a point which has important ethical implications let us review the evidence once again. The question we must address is whether or not every fertilised ovum is animated by the consciousness of an intermediate being. An answer in the negative will mean that the Buddha believed that the genetic basis for individual life could exist independently of an animating consciousness. The first two conditions, which the Buddha tells us can be fulfilled without the third, require that intercourse should take place and that the woman be in her fertile period. In terms of the embryology of the time, intercourse was understood as a process leading to the mingling of semen and menstrual blood. The question then is: can this ancient notion of the mingling of the male and female contributions to the physical substrate of the new conceptus be taken as equivalent to the modern understanding of conception as the fusion of sperm and ovum?

We have assumed above that it can, and there seems to be nothing in Buddhist doctrine which would lead us to revise this opinion. The Buddha taught that human nature was a composite of spiritual and physical elements, and in terms of the contemporary understanding of procreation the biological aspect of human life was generated through intercourse in the manner described. No further *biological* conditions over and above intercourse in due season are stipulated for generation to take place. It may be objected that we suggested earlier that when the spiritual and physical aspects arise they always do so together, but now we are allowing that the material basis for life can arise on its own. Is there not a contradiction here? Not necessarily, for what we suggested earlier was not that the two are always conjoined, but that the two always arise simultaneously *in the generation of a new individual*. In the present instance no new individual life has come into being, only the biological basis for one.

What would be the future of such an unanimated conceptus? From the perspective of Buddhist doctrine it would seem impossible for it to develop very far. We noted above a statement by the Buddha that normal embryonic development could not continue in the absence of the psychic element in the mind-body aggregate. He stated that if consciousness (*viññana*) were 'extirpated' from one still young, then normal growth and development could not continue. This suggests that *viññana* in some sense orchestrates the evolution of the physical body. How this happens is unclear and we can only speculate. In our example of the computer in Chapter 1 we compared consciousness to the electric current which flows through the components. The image of an electric light has also been used. Francis Story has suggested that at conception 'The released energy [*viññana*] in some way operates on and through the combination of male and female generative cells on much the same principle as that of the electric current working on the filaments in the lamp to produce light.'[38] Again, we might picture the consciousness of the intermediate being engulfing the fertilised ovum in a psychic field, rather like a magnetic field surrounding a magnet. The presence of this field might be required for the development of a human embryo beyond a very early stage. An embryo which was not animated would lack this ability to evolve: most probably it would develop abnormally and be lost in the course of the menstrual cycle.

Summary

We summarised above the understanding of conception and embryology found in the early sources. Having considered this in the light of modern knowledge, what modifications are required? Although the three conditions announced by the Buddha are still valid, we suggest that if he were making his statement today he would make reference to two conditions rather than three. This would be done by collapsing the first and second. Buddhism would then hold that conception takes place when (i) the ovum is fertilised by the sperm; and (ii) an intermediate being is available to be reborn. These two conditions would apply to conception in the normal manner through sexual intercourse and also to *in vitro* fertilisation, where conception takes place in the laboratory. If we desire further precision on the precise moment of fertilisation, we must enquire whether conception occurs on the penetration of the outer layer (*zona pellucida*) of the ovum or when the two sets of twenty-three chromosomes fuse together (syngamy). The earlier of these occurs about two hours after intercourse and the latter around twenty hours later.[39] Although this is a fine distinction to make, if we are to choose one or the other there are grounds for regarding the earlier event, sperm penetration, as the point of origin of the new individual. When penetration has occurred, and the sperm releases its genetic components, all the ingredients are together within a single cell. No new genetic information will be added from this point on, and syngamy can itself be seen as an event in the unfolding of this natural development.

III   EMBRYO LOSS AND TWINNING

Having updated the Buddhist view as to when individual life begins, we now consider some scientific evidence which may be thought to challenge it. We have suggested that Buddhism holds the view that individual life begins at fertilisation. This is by no means an original suggestion and is widely understood to be the Buddhist position, although it has not so far been subject to critical examination. Important moral conclusions flow from this belief, but we are not concerned with those matters at this stage and will discuss them in due course below. For the time being we are concerned only with the defensibility of the claim itself. Modern research has made a number of discoveries which seem to challenge it, and which have impressed some as being powerful arguments against it. Two, in particular, are of importance: the first concerns the natural loss of fertilised embryos, and the second the phenomena of twinning and recombination.

Loss of fertilised embryos

Research has shown that a high proportion of fertilised embryos are lost prior to and during implantation. Although estimates of the numbers lost vary widely, all that need be acknowledged is that a significant proportion, perhaps the majority, of pre-implantation embryos do not survive. The majority of the losses seems to be attributable to chromosomal and genetic defects. It has been suggested that the discovery of embryo loss poses a problem for the view that life begins at fertilisation, since it would mean that many individuals die at a very early stage of their lives. While these statistics do not show that life *cannot* begin at fertilisation, they may be inconvenient in relation to other beliefs. In a Christian context, for example, they may seem incompatible with a belief in immediate animation (although this has never been asserted as a dogma) since it would mean that God had chosen a very inefficient method of delivery for the human lives which he creates. Since Christianity maintains that each life is once and for all, there would be millions of souls who had been denied the chance to live meaningful lives.

As a general response it is open to both Buddhists and Christians to claim that the statistics concerning embryo loss carry very little weight as an argument against life beginning at fertilisation. The fact that many individuals die at an early point in their lives, it could be said, is not in principle different from them dying at some later stage. Everyone dies at some time or another, and for most of human history the infant mortality rate has been as high as 50 per cent.[40] The significance of human existence, however, is surely not to be correlated with some minimum period of life on earth. Although unwelcome to us, these statistics do little more than confirm the fragility of human life, a fact of which both Buddhism and Christianity are only too aware.

Embryo loss and Buddhism

There are further reasons why embryo loss has even less significance for Buddhism. In the first place, souls are not created by God so the problem of a wise God choosing a precarious method of delivery for his created souls does not arise. According to the doctrine of karma, it is individuals themselves who are responsible for the circumstances of their birth. Second, since salvation does not depend upon a single lifetime it is of no great consequence even if the fertilised embryo chosen as the vehicle for rebirth fails to implant and develop; another opportunity would soon arise, perhaps shortly after with the same parents.

Early Buddhists were quite aware that embryos died in the womb at all stages of pregnancy.[41] Indeed, Buddhaghosa goes so far as to suggest that infertility is never due to a failure to conceive but to the death of the early embryo from two karmically-conditioned causes: either the effect of bodily 'humours' or attack by micro-organisms.

A 'barren woman' means one who cannot conceive. But in fact there is really no such thing as a woman who cannot conceive, and this phrase is actually used of a woman who *does* conceive but in whose womb the embryo fails to become properly established. It is thought that all women conceive during their fertile period, but in the case of the one described here as 'barren' there occurs a maturation of bad karma for the beings who enter her womb to take rebirth. They, who have taken rebirth through the maturation of only limited virtue, are overcome by the maturation of the evil karma and die. In a new conception there are two reasons why, in accordance with its karma, an embryo does not become established: [the bodily humour of] wind, and the [action] of other organisms. Dried up[42] by the wind the embryo is made to disappear, or else it is consumed by organisms.[43]

On the above understanding embryo loss could occur on a scale equal to anything suggested by modern research.

A further response to embryo loss could be made by Buddhism. We suggested above that Buddhism is not committed to the view that every fertilised embryo is animated. Instead, fertilised ova may best be seen as opportunities for the intermediate being to take rebirth. The fact that many fertilised embryos were lost, therefore, need not mean that an equal number of individuals had died. We may enquire as to *why* the third condition is not fulfilled in all cases of fertilisation with the result that some embryos remain unanimated. Based on the Buddha's statement, the logical explanation would be that there is no intermediate being available to take rebirth at the time. Another explanation which is consistent with traditional views might be that those beings which *are* available to be reborn are karmically unsuited to the present couple as parents. A third possibility might be to turn the matter around and ask if the high rate of embryo loss can itself be *explained* by reference to the Buddhist account of conception. Perhaps many of the lost embryos are lost precisely *because* the third condition has not been fulfilled. It may be, for example, that certain ova have genetic defects which prevent the intermediate being from merging with them at fertilisation. In the absence of an animating consciousness, these fertilised ova would then lack something essential to their successful growth and development. One obstacle in the way of this hypothesis is the fact that many people are born with genetic defects, so that any mechanism which operates in the way suggested is less than perfect. Further conjecture in this direction would take us away from our main theme, and for the present we will simply note that embryo loss poses no real problem to the Buddhist belief that individual life *when* it begins, begins at fertilisation. The phenomenon of embryo loss does not show that the embryos which are lost cannot be human individuals. The statistics quoted in this connection are simply irrelevant one way or the other to the question of when life begins.

Twinning

A more serious objection to the view that life begins at fertilisation arises from the discovery of twinning and recombination in the early embryo. In rare instances the early embryo can split and develop as two (or more) genetically identical but separate units. Each cell is able to develop into an adult human being and each possesses the same potential as the original fertilised egg. This is the process which leads to the birth of identical twins. Identical twins can trace their origins back to the same fertilised embryo or zygote, and are accordingly referred to as *monozygotic* twins. An even rarer occurrence is that after splitting, the separate cells may recombine and continue to develop normally as a single being. The limit for these developments is reached around the end of the process of implantation, some fourteen days after fertilisation. It is at this time that a small line can be observed within the embryo known as the 'primitive streak'. This coincides with the appearance of the central nervous system. The conclusion drawn by some commentators from these facts is that for the first fourteen days of its life the early embryo cannot be regarded as an ontological individual.[44] If it is not an ontological individual, the argument goes, then no human individual can be present. The early embryo may be in the *process* of becoming a human individual ('hominisation'), but it has not yet become one. Since our moral obligations are towards human individuals, we find ourselves in the presence of something which lacks moral status. The entity we have before us might best be regarded as a biological matrix out of which individual human life or *lives* may emerge.

Twinning and psychic splitting

The question which now arises is how Buddhism would explain
these scientific facts. In the case of twinning we clearly end up with
two distinct individuals from the original unicellular fertilised
zygote. If, as we maintain, individual life begins at fertilisation, how
is the appearance of a new individual later than fertilisation to be
explained in terms of Buddhist doctrine? Could it be that the original individual has divided into two? If such is the explanation it
would require the conclusion that the animating intelligence
(*viññana*) of the intermediate being which took rebirth has divided
into two parts, with one part in each twin. David Stott, writing from
the perspective of Tibetan Buddhism, suggests that such is precisely
what has happened:

> Therefore, although one talks of mind 'entering the unified sperm
> and ovum', and consciousness, or sentience, 'pervading all forms
> of life' since mind's intrinsic nature is emptiness, it cannot be
> found or located within it as if it were some kind of supernatural
> substance. Since it is intrinsically empty, having no attributes such
> as form or shape, even if early cellular development led to a split-
> ting into two as occurs with identical twins, the mental energy
> would itself split to 'pervade' the two newly distinct embryos.
> Since mental energy is intrinsically empty it follows that it can
> both split and recombine ad infinitum.[45]

This explanation, however, leads to conclusions which do not sit
well with Buddhist views on personal identity. If the consciousness
of the intermediate being has split, as suggested above, we would
then have two beings in existence who were identical both psychi-
cally (*nama*) and genetically (*rupa*). Both would, in theory, be able to
look back into their previous lives where they would find an identical biography. Again, if there could be two such beings, there could
in principle, given the power of sentience to split ad infinitum, be
three, four or a hundred beings all identical in every way. Paradoxically, although several beings would now have identical recollec-
tions, only one would ever have 'lived' in any real sense as the
embodied being who actually did the things remembered.

A modification to Stott's explanation would be to accept that
viññana did in fact split, but that only one part was heir to the historical continuum of the original being. This would mean that the

original intermediate being had animated its genetic twin as if by osmosis, without transferring its own unique karmic identity to it in the process. However, this explanation would then create a further problem since it would mean that at a defined moment a new individual being had sprung into existence. The orthodox Buddhist view of these matters is that individual existence has no beginning, and I think it would be hard to find canonical authority for the view that a karmic individual can be originated in a way such as this, or indeed in any way. There would also be the problem of the karmic status of the newly originated individual, or more accurately, its lack of one. In terms of Buddhist psychology it is difficult to conceive of a being as a karmic *tabula rasa*, a fact which makes this suggestion, too, seem implausible. It looks, therefore, as if the conclusions which follow from the postulate of *viññana* splitting as an explanation of twinning seem at variance with basic Buddhist views on personal identity. Although Stott's suggestion is intriguing, and his paper received the approval of two distinguished Tibetan lamas, it raises problems about personal identity which seem difficult to resolve. For the present it is proposed to keep to what we understand to be the orthodox view, namely that every individual is an historical being with a unique karmic biography. It follows from this that no two individuals can be both genetically and karmically identical. We have Buddhaghosa's authority for the fact that no two individuals, even twins, are completely identical.

> Even when the external circumstances [of rebirth] are the same, it can be seen there is a difference with respect to [moral] inferiority and superiority, and so forth. Even when the external circumstances such as father, mother, sperm, blood, and nourishment are the same, a difference can still be seen in the case of twins with respect to [moral] inferiority and superiority, and so forth.[46]

Elsewhere, he suggests that subtle behavioural differences will reveal themselves even in identical twins.

> Amongst the infinite human beings in the infinite universes no two are exactly alike in complexion, appearance and so forth. Even twins who are the same in complexion and appearance will be different in the way they look forward or backward, speak, laugh, walk, stand and so forth, hence they are said to be 'different in body'.[47]

Alternatives to psychic splitting

If the hypothesis of psychic splitting is rejected, we are led to the conclusion that in the case of twinning each of the twins must be the embodied form of separate intermediate beings. The problem still remains, however, that one of these individuals was seen to come into being by emerging from the other after fertilisation. This seems to cast doubt on the notion that fertilisation marks the beginning of individual life. A further puzzle is to explain what takes place in recombination. Apparently the one individual who became two has now become one again! To add to our bewilderment in the face of these discoveries, it has been suggested that these developments are not confined to exceptional cases such as identical twins, but that *every* embryo has this potential for twinning and recombination and these phenomena can be reproduced by artificial means *in vitro* in the laboratory. In the light of these remarkable facts perhaps it is no wonder that some have felt it safer to reserve judgement altogether on the question of when individual life begins.

On further reflection, however, there are grounds for thinking that the significance of this evidence (as with natural embryo loss) has been overstated. Before proceeding further it would be well to recognise both the rarity of twinning in practice (around three to four cases per thousand births)[48] and the uncertainty among scientists as to the underlying causes. The division or aggregation of embryonic cells is only possible under restricted conditions and not at all phases of development. The mechanism by which twinning occurs is as yet imperfectly understood, as is the process of fusion or recombination. Artificially induced twinning is possible in some non-human species but not in others, and it has not yet been established whether recombination occurs naturally in humans or other mammals. In short the empirical data is inconclusive, and to attach great significance to conclusions reached on the basis of it would be inadvisable. Notwithstanding the above caveat, however, we may proceed to enquire how the data as presently understood could be interpreted in accordance with Buddhist views on the origin of individual life.

Conceptual considerations

The problems in understanding these phenomena are not purely of a scientific kind. To be endowed with meaning, the scientific data needs to be located within a conceptual framework. This is a philosophical exercise, and the conclusions reached concerning twinning will depend as much upon the conceptual mapping of the data as the 'facts' themselves. In favour of the view of the pre-implantation zygote as an ontological individual is the fact that definite criteria can be given which identify each one uniquely.[49] For example, each is visibly identifiable as a separate and unified entity, developing in a self-directed and self-organising way in accordance with its unique genetic constitution. These factors make it distinguishable from all other human beings, including its own parents. There is thus a strong presumption in favour of ontological individuality. But if the embryo is an ontological individual, how is twinning to be explained? We suggest that monozygotic twinning, which occurs with the first division of the unicellular zygote, should be thought of as a process whereby the zygote produces a genetic duplicate of itself. In so doing, the original zygote retains its individuality, while creating another individual through mitotic cleavage. Both embryos then develop separately to produce genetically identical adults.

Sexual and asexual reproduction

What has been shown by twinning, on this alternative account, is that a new being has come into existence at a point shortly after fertilisation. There is no reason why this need present a problem to the Buddhist account of the circumstances under which life begins. It is not unreasonable to assume that when the Buddha described the three conditions necessary for conception he was describing the *normal* process of human sexual generation. In other words, the Buddha was not saying there is only *one* way in which life can come into being, and his statement need not be read as excluding other possibilities. One reason for thinking this interpretation is correct is that Buddhism recognises a variety of ways in which life can be produced. A stock list of four 'wombs' (*yoni*) is commonly given in the texts: thus life can originate from an egg, from a womb, from putrefying matter, and by 'spontaneous generation'.[50] Birds and reptiles proceed from the first, mammals from the second, and insects (as Aristotle also believed) from the third. As regards the last, it was thought that sages and some supernatural beings have the power to materialise a human form for themselves at will. We may suppose that if this happens at all it is very rare. What it suggests, however, is that although sexual reproduction is overwhelmingly the most common way in which human life begins it was not thought of as excluding other possibilities.[51]

Apart from cellular division there may be other ways in which new individual life can be created asexually, for instance by cloning from adults. A technique such as this, if it is ever perfected in human beings, would show only that there are a variety of ways in which life can be generated. It would not cast doubt on whether the host from which the clone was taken, or the clone itself, were ontological individuals. It is known that amoebas and plants can produce duplicates of themselves, but no one would suggest that because a rose-bush has the capacity to produce a clone it is not an individual plant itself.

In the present context it can be argued that what has been shown by twinning is that in addition to the normal mode of sexual reproduction referred to by the Buddha there are also modes of *asexual* reproduction, of which twinning is one example. This mode of asexual reproduction is in fact precisely how every normal embryo grows. Through the process of mitotic cleavage the cells multiply from one to two, two to four, four to eight and so on, until the embryo consists of over a hundred cells by the time of implantation. Of course, it is only in very rare cases that this process of duplication results in the formation of a second embryo. What has happened in these cases is that sexual reproduction through coitus has been replaced by another kind of generation involving the division of cells. From the Buddhist perspective, it could be said that the fertilised ovum has produced a clone which was then separately animated. In terms of Buddhist philosophy each of these embryos would be a new individual life informed at the time of its creation by the spirit of an intermediate being – one at fertilisation and the other at the time of twinning.

Recombination

Although twinning is rare, recombination is rarer still. There is some evidence that it can happen in the case of certain animals, but it does not happen naturally in humans. However, let us grant for the sake of argument that it may occur in humans and that two separate embryos can fuse together as one. What, then, has taken place? It seems unlikely that Buddhism would accept where human beings are concerned there could be two *viññanas* sharing the one physical substrate, so a fusion of two beings must be ruled out. The next question to consider is whether the second embryo was animated or not. Let us assume, first, that it was not. Recombination in these circumstances would be of little significance: the ability to assimilate cells in this way is an interesting property of the early embryo, but it is of no moral significance. It does not show that that an embryo is not an individual any more than when a man eats an oyster it proves that the oyster was not an individual because it has now been assimilated into another organism. Turning to the remaining possibility, if the second embryo *was* animated the explanation for what has occurred must be that it died and the first embryo, as before, re-assimilated its cells.

The conclusion Buddhists may wish to draw from the above discussion is that individuals can begin their lives in many ways. In the overwhelming majority of cases individual life is generated through sexual reproduction and begins at fertilisation. In a tiny minority of cases, such as occurs with identical twins, life may come into being slightly later than fertilisation through the process of asexual reproduction known as twinning. There is some evidence that this process can be stimulated artificially in the laboratory, but it is of no greater significance when it occurs artificially than when it occurs spontaneously. At some time in the future individuals may begin their lives by cloning, and perhaps by other means as yet unknown. From an ethical perspective the Buddhist position would be that once we have an individual before us (and the presumption must be that we always have *at least one*),[52] then that being is entitled to full moral respect and protection, however long or short its life and regardless of when and by what precise means its physical nature came into being.

Up to this point our efforts have been directed to clarifying the Buddhist understanding of the reproductive process, and to some extent defending it in the face of evidence which might seem to challenge it. Now that we have clarified the traditional Buddhist understanding of how and when individual human life begins we can proceed to generate a Buddhist response to some of the major biomedical issues which arise at the beginning of life.

IV ABORTION

References to abortion in this context concern deliberately induced or therapeutic abortion as opposed to spontaneous abortion or miscarriage. It is impossible to tell from the early sources how wide-spread the practice of abortion was at the time of the Buddha. Pre-Buddhist Indian sources dating from as early as 1200 BC condemn abortion and stress the moral inviolability of the fetus.[53] Following a review of the textual evidence on abortion in Hinduism Lipner sums up as follows:

> From this we conclude that the unborn, in classical Hindu tradition, were accorded a moral status deserving of special protection and that abortion was generally reprehensible because thereby the integrity of the human person (of both victim and abortionist) was seriously violated.[54]

Despite the widespread condemnation of the practice, however, abortions were sought. In ancient India, as in many societies, religious practitioners adopted the roles of counsellors, astrologers and physicians, and were consulted by their clientele on a variety of matters of which the age-old problem of an unwanted pregnancy was one. The Buddha discouraged monks from engaging in this type of activity, and in one of his early sermons it is stated that a good Buddhist monk, unlike members of certain other sects, does not dabble in superstitious practices designed to bring about results such as abortions.[55] None the less, it is clear from the case histories in the Monastic Rule we shall consider below that monks were consulted about such things and did on occasion become illegally involved.

Reasons for seeking abortions

The reasons for seeking abortions were varied. They included concealing extramarital affairs, preventing inheritances, and domestic rivalry between co-wives. There appear to be no examples in Buddhist literature of abortion performed for medical reasons.[56] The methods used to procure abortions included ointments, potions and charms, pressing or crushing the womb and scorching or heating it. We are told in the *Jataka* that the queen of king Bimbisara, a contemporary of the Buddha, resorted to having her womb massaged and heated by steam in order to cause the death of the child she was carrying.[57] In the event the child survived and eventually murdered his father. Another story, from the *Dhammapada* commentary, tells of the rivalry between co-wives. The elder wife, who was barren, tricked the junior into repeatedly aborting in order to preserve her seniority in the family. The story goes on to describe the evil consequences which pursued the elder wife in future lives.[58]

In societies everywhere a great deal can turn on the birth of a child, such as his or her becoming heir to a kingdom or other property, and there will always be rival interests which do not have the welfare of the child at heart. In his commentary on the Monastic Rule, Buddhaghosa refers to a passage from a *sutta* describing how magic powers might be used to thwart a normal delivery:

> Moreover, monks, a religious wanderer (*samana*) or a brahman who has achieved psychic control and mental mastery may direct evil thoughts towards the embryo in the womb of some woman with the wish that the embryo in the womb should not be delivered safely. In this way, monks, there is the slaying of [the heir to] an estate.[59]

The prohibition on abortion in the Monastic Rule

Abortion is prohibited by the precept against 'depriving a human being of life'. That cases of abortion were classified under this rubric suggests that causing the death of a fetus was considered as grave an offence as killing an adult. The precept states that it is wrong to 'intentionally deprive a human being of life', but is a fetus a human being? The precept is also found in the following version, which answers this question and makes clear that the life of a fetus is entitled to protection under the precept:

> An ordained monk should not intentionally deprive a living thing of life even if it is only an ant. A monk who deliberately deprives a human being of life, even to the extent of causing an abortion, is no longer a follower of the Buddha. As a flat stone broken asunder cannot be put back together again, a monk who deliberately deprives a human being of life is no longer a follower of the Buddha. This is something not to be done by you as long as life lasts.[60]

The Monastic Rule also provides helpful explanations of some of the terminology used. The term 'human being' is defined as follows:

> A human being [exists] in the interval between the first moment when mind arises in the mother's womb [that is to say] the first manifestation of consciousness (*viññana*), and death.[61]

This suggests that a human being comes into existence at conception. Buddhaghosa expands on this as follows:

> Now in order to show that the phrase *should deprive a human being of life* refers to human nature (*manussata-bhava*) from the very beginning onwards, he begins the passage with the words *a human being*. [The phrase] *in the mother's womb* is used to draw attention to the extremely delicate mode of being (*attabhava*) of those who have entered the womb. *The first moment when mind arises* means 'the first moment when mind arises in the new existence'. *Arises* means 'is born', and the *first manifestation of consciousness* is another way of saying the same thing. By the phrase *the first moment when mind arises in the mother's womb* is shown the complete (*sakala*) reinstatement of the five categories [of human nature]. So the very first moment of existence in human form consists of that first moment of mind, with its three associated immaterial components [i.e. feeling, thought, and character], and the [material] body (*rupa*) of the embryo (*kalala*) which is generated along with it.

The passage goes on to make clear that the offence applies to the destruction of human life at any time between conception and death:

> The individual being (*attabhava*) begins from this tiny substance [and] gradually grows old with a natural lifespan of up to one hundred and twenty years. Throughout all of this until death, such is a human being. [The phrase] *who should deprive it of life* means 'separating from life' either at the stage of the embryo (*kalala*) by scorching, crushing, or the use of medicine, or at any subsequent stage by some similar kind of assault.[62]

The above canonical and commentarial evidence from the Monastic Rule is consistent with the conclusions we reached in our discussion of conception. We saw there that human life was thought to begin at fertilisation, and it is logical, therefore, that the offence of depriving a human being of life should apply from this point onwards. Buddhaghosa makes explicit reference above to the earliest stage of the embryo (*kalala*) and to the techniques of scorching and crushing which were used to destroy intra-uterine life.

Cases of abortion in the Monastic Rule

Let us now consider the details of the cases of abortion recorded in the Monastic Rule.[63] There are seven cases in all. The first concerns a woman who became pregnant by a lover while her husband was away from home. She appealed to a monk who was close to the family to bring her an abortive preparation. He gave this to her and the child died. The second case involved domestic rivalry between two wives in the manner of the incident from the *Dhammapada* commentary referred to earlier. The barren wife asked a monk to administer a preparation to her fertile co-wife so that the latter would not be favoured and allowed to rule the roost. The monk agreed and the child died. The circumstances of the third, fourth and fifth cases are identical but the outcomes vary. In the third case the mother dies, in the fourth case both mother and child die, and in the fifth case neither die. In the sixth case a monk advises a woman that the child in her womb can be killed by crushing, and in the seventh case by scorching. In both cases the result was the death of the child.

In all of the cases where the abortion brings about the death of the child as intended, the judicial decision was that the offence fell into the category of 'depriving a human being of life'. There was no reduction in the gravity of the offence by virtue of the fact that the victim was a child *in utero* as opposed to a child already born or an adult. In forensic terms no significance seems to have been attached to the particular gestational phase of the fetus when the abortion occurred, and there is no indication in the text or in the commentary that the offence became graver as the fetus approached full term. We have seen that early Buddhism distinguished several stages of embryonic development, but there is no evidence that this classification was relevant from a moral or legal perspective.

As a final piece of evidence from the Monastic Rule we may cite a comment made by Buddhaghosa in connection with his discussion of killing by means of digging a pit as a trap. What, he asks, would be the position if a pregnant woman fell into the pit? Would this count as one offence or two? If it counted as one we could conclude that the life of the fetus was not regarded as equal in value to that of an adult. And what if the mother survived but the child died? If the verdict was that the death of the child was also a breach of the rule it would confirm the evidence seen thus far that no moral or legal distinction is made between killing a born child and a child in the womb. Fortunately, Buddhaghosa is explicit about the decision across the range of possible outcomes:

> If a pregnant woman falls in and dies along with her child (*sagabbha*), this counts as two breaches of the precept against taking life (*panatipata*). If the child [alone] dies there is one [breach], and if the child does not die but the mother dies there is also one.[64]

The victim and the gravity of the offence

There is evidence in some sources to suggest that the gravity of an offence was thought to vary in accordance with the nature of the victim. The religious status of the victim is the feature most frequently mentioned: it is said, for example, that it is worse to injure a Buddha than an ordinary person. Another factor sometimes mentioned is the victim's physical size. This factor is sometimes thought to apply to abortion, such that the offence becomes worse the more advanced the gestational stage. Peter Harvey, for example, reports that 'The bad karma of an abortion is said to vary according to the size of the foetus'.[65] Trevor Ling also found evidence of this view among Buddhists:

> In general it can be said that in Theravada Buddhist countries the moral stigma which attaches to abortion increases with the size of the foetus. This is an aspect of the general Buddhist notion that the seriousness of the act of taking life increases with the size, complexity and even sanctity of the being whose life is taken. It is relatively less serious to destroy a mosquito than a dog; less serious to destroy a dog than an elephant; it is more serious to take the life of a man than of an elephant, and most serious of all to take the life of a monk. It would thus be less serious to terminate the life of a month-old foetus than of a child about to be born.[66]

The principles involved here are certainly ancient, although we suggest they were never intended to be applied to abortion. To demonstrate this, however, will involve a short detour from the issue of abortion itself.

Sanctity

In the passage above, Ling has identified three scales of values involved in assessing the status of a victim: size, complexity, and sanctity. We will consider the last one first. According to this criterion, the gravity of the offence of taking life is greater the more spiritually advanced the person killed. We concluded at the end of Chapter 1 that while the intentional destruction of karmic life is always morally wrong, the value of different forms of life can vary. It follows that while intentional killing is always wrong without exception,[67] not every case of it is equally grave and the karmic consequences need not be identical. We might draw an analogy with the law of theft: while all theft is a crime, the theft of one pound will be punished more leniently than the theft of a thousand pounds.

But what is it that makes the wrong more profound when a Buddha is killed as opposed to an ordinary person? A first thought might be it is because the life of a Buddha is more valuable. The *life* of a Buddha, however, in the sense we spoke of life in Chapter 1, is neither more nor less valuable than the life of anyone else. What makes a Buddha distinctive cannot be explained in terms of any *one* basic good, and the great 'sanctity' of a Buddha is due to his participation in *all* the goods open to human realisation. A Buddha is a living celebration of human potential. To kill a Buddha would be to destroy not only life, but also the other goods such as knowledge and friendship which he has fulfilled to perfection.

Size and complexity

Size and/or complexity are the two other factors to which reference was made. Some care is required here, for these two measures can easily be confused. The sequence of mosquito–dog–elephant–man mentioned by Ling seems to do precisely that, in that the two distinct sets of criteria have become mixed together. This hierarchy cannot be explained consistently either by reference to size or complexity. In terms of size we begin with the mosquito and should logically end with the elephant: instead we find man at the top, who is obviously smaller than an elephant. It is clear, however, that size cannot outweigh complexity when measuring the moral gravity of an offence, for if it did it would be more serious to kill an elephant than a man. And as regards complexity, does it makes sense to say

Buddhism & Bioethics

that an elephant is more 'complex' than a dog? Isn't the difference simply that an elephant is *bigger* than a dog? To make any sense of these criteria, size would have to be seen as a subcategory of complexity. In the example above, then, there would really be three examples of 'complexity' rather than four. In terms of complexity the hierarchy would be one of insect, animal, and human life. The dog and the elephant are both examples of animal life and hence equal in complexity. Where they differ could be explained in terms of the subcategory of 'size'.

Buddhaghosa on size and sanctity

There is some doubt, however, whether 'complexity' is really a separate criterion at all for Buddhism, and it may be that Ling's reference to it is only intended as a gloss on 'size'. Buddhaghosa, for example, does not mention complexity in his explanation of how the gravity of offences varies with the victim, and confines himself instead to the twin criteria of size and sanctity.

> Taking life in the case of [beings such as] animals and so forth which are without virtue (*gunavirahita*) is a minor sin if they are small and a great sin if they are large. Why? Because of the greater effort required. In cases where the effort is identical, the offence may be worse due to greater size. Among [beings such as] humans and so forth who have virtue (*gunavant*), it is a minor sin to kill a being of small virtue but a great sin to kill a being of great virtue. Where both bodily size and virtue are the same, it is a minor sin if the wickedness (*kilesa*) involved and the assault itself are moderate, and a great sin if they are extreme.[68]

This seems to confirm our suggestion that there are really only two criteria, size and sanctity. But why should size be of any importance at all from a moral perspective? Why is it that killing an elephant is worse than killing a dog? Buddhaghosa supplies a partial answer when he tells us it is 'Because of the greater effort required.' Thus killing an elephant would be more serious than killing a dog, because while a dog could be killed with a single blow, to kill an elephant would probably be a long and bloody affair. It would involve planning and organisation, and probably repeated assaults upon the creature with primitive weapons before it could be killed. The rationale from an ethical point of view thus turns out not to be the size of the creature at all. Size here is only shorthand for the determination on the part of the assailants to do wrong. It is their premeditated, calculating and wilful perseverance which makes the act worse than it would otherwise be. This does not mean that the death of a smaller creature cannot be planned and executed with the same cruel intention, but the size of the victim would normally correspond to some degree with the strength of the murderous intent. Tiny creatures can sometimes be killed absentmindedly, for example when we step on an ant. Death in such cases is unpremeditated. Size can thus be seen as correlative to the degree of premeditation when life is destroyed.

The mention of 'effort' in Buddhaghosa's statement allows us to make sense of the matter in the way above. However, he then goes on to complicate the picture by adding 'In cases where the effort is *identical*, the offence may be worse due to greater size.' This suggests that if we could kill either a dog or an elephant with the same effort, for example with a single shot from a rifle, it would still be more serious to kill the elephant. Unfortunately, he offers no explanation for this. Perhaps size is appealed to in this rather unconvincing way as a factor which, *faut de mieux*, might be used to separate two otherwise identical cases for forensic purposes, just as the gravity of murder among humans can be distinguished according to the sanctity of the victim. If this is not the explanation it is difficult to see what Buddhaghosa was getting at, and we must let the matter rest there.

Returning to the subject of abortion, however, the important point to note is that Buddhaghosa's discussion of size does not relate to *human beings* in any way. He states quite clearly: 'Taking life in the case of [beings such as] *animals* ... is a minor sin if they are small and a great sin if they are large.' Only after discussing considerations of size and effort does he turn to the case of humans and say: 'Among [beings such as] *humans* ... it is a minor sin to kill a being of small virtue but a great sin to kill a being of great virtue.' The two categories of humans and animals, and the related criteria, are thus quite separate in his mind. What this indicates is that of the two criteria *sanctity* is the only one which has any relevance to human victims. Should the need arise to distinguish between two cases where the victims are of equal sanctity, the degree of violence in the assault can be considered as an aggravating factor, as can size in the case of the killing of two animals. The key point to grasp, however, is that size is a relevant factor in assessing the gravity of a breach of the first precept *only when the victim is an animal*.

Non-textual considerations

Apart from textual considerations, logic drives us to the same conclusion. If the criterion of size is to be applied to human beings it must surely be done so consistently, as it is with dogs and elephants. This means that the criterion cannot be restricted to the case of a child in the womb. In the example it is applied to dogs and elephants who are born, not dogs and elephants *in utero*. Applying this principle logically, however, would lead us to conclude that it is less serious to kill a two-year-old child than a five-year-old. Likewise, it would be less serious to kill an eight-year-old than a teenager, and less serious to kill a teenager than an adult. In general it would be less serious to kill women than men. If size is of any importance at all, the worst cases of murder would be those of Sumo wrestlers. We can see from this that the entire line of argument from size to moral seriousness is ridiculous when applied to human beings. We can deduce, furthermore, that the criterion of size is not intended to apply between members of any *one* animal species. The criterion only makes sense when applied *across* species in the animal kingdom. Thus it would be valid when contrasting a small species (dogs) with a large species (elephants), but not when contrasting small members of the same species with larger members (e.g. small elephants with large elephants).

We saw that in his commentary on the offence of taking human life in the Monastic Rule, where abortion is discussed at length, Buddhaghosa nowhere mentions the size of the fetus as having any bearing on the gravity of the offence. The view that the size of the fetus has any moral relevance to abortion, then, seems to be the result of a misapplied criterion. This may be due to genuine confusion about the way Buddhism modulates the gravity of offences by reference to the status of the victim. On the other hand it could be an example of *bricolage* (of which more below) whereby an ambiguity has been creatively exploited as a conscience-salver.

Situation ethics and Buddhism

In spite of the clear condemnation of abortion in the ancient sources, some contemporary writers have adopted an alternative and more liberal approach to the traditional views described above. In a recent book on socially engaged Buddhism, Ken Jones cites examples of 'situational morality at work', one of which relates to abortion. He

sees abortion as a dilemma for Buddhism, given that it is both profoundly compassionate and yet counsels against all killing. This
raises issues of two kinds which will be discussed in turn. The first
concerns the merits of the situational perspective itself; the second
picks up an issue raised in Chapter 1 concerning what can legitimately be advanced as a 'Buddhist view'.

In *The Social Face of Buddhism*, three widely quoted opinions on
abortion are cited, two from (Western) Zen Masters and one from
the Shin Buddhist Churches of America Social Issues Committee.
We will list all three, beginning with the last:

> Although others may be involved in the decision-making, it is the
> woman carrying the fetus, and no-one else, who must in the end
> make this most difficult decision and live with it for the rest of her
> life. As Buddhists we can only encourage her to make a decision
> that is both thoughtful and compassionate.[69]

Zen Master Philip Kapleau advises:

> If your mind is free of fear and of narrow selfish concerns, you
> will know what course of action to take. Put yourself deeply into
> *zazen* [meditation] – look into your own heart-mind, reflecting
> carefully on all aspects of your life situation and on the repercus-
> sions your actions might have on your family and on society as a
> whole. Once the upper levels of mind, which weigh and analyse,
> have come to rest, the 'right' course of action will become clear.
> And when such action is accompanied by a feeling of inner peace,
> you may be sure you have not gone astray.[70]

Zen Master Robert Aitken comments:

> Sitting in on sharing meetings in the Diamond Sangha, our Zen
> Buddhist society in Hawaii, I get the impression that when a
> woman is sensitive to her feelings, she is conscious that abor-
> tion is killing a part of herself and terminating the ancient
> process, begun anew within herself, of bringing life into being.
> Thus she is likely to feel acutely miserable after making a
> decision to have an abortion. This is a time for compassion for
> the woman, and for her to be compassionate with herself and
> for her unborn child. If I am consulted, and we explore the
> options carefully and I learn that the decision is definite, I

encourage her to go through the act with the consciousness of a mother who holds her dying child in her arms, lovingly nurturing it as it passes from life. Sorrow and suffering from the nature of samsara [the transitory world of phenomena], the flow of life and death, and the decision to prevent birth is made on balance with other elements of suffering. Once the decision is made, there is no blame, but rather acknowledgement that sadness pervades the whole universe, and this bit of life goes with our deepest love.[71]

Representative opinion

In citing the three views above Ken Jones invites us to take them as representative of the views of Western Buddhists. 'The three statements by Buddhists in the West which I have encountered on this topic,' he writes, 'all come to similar conclusions.'[72] It must be pointed out, however, that the sample of opinion offered is a limited one and a more representative range of Buddhist views might have been quoted. David Stott, for example, has expressed the following viewpoint on abortion:

> In this respect, it is vital to realise that it is the taking of life from the very moment of conception until the moment of death, which is forbidden by Buddha. Thus the performance of abortion or fatality-causing experiments on the unborn child constitute the taking of life, just as surely as the taking of life at any other point in the continuum of conception to death.[73]

And in an article entitled 'A Buddhist View of Abortion' Phillip Lecso concludes: 'Buddhism is firmly against abortion ... Buddhism rejects the arguments favouring abortion and argues strongly for protecting all human life'.[74] Finally, in canvassing a representative range of opinion on the matter it would not have been unreasonable to turn to the *Encyclopaedia of Buddhism*, where under the entry on 'Abortion' we read:

> Hence, there is no doubt about the unequivocal attitude of the Buddha's teaching in respect of life from the very inception of conception, i.e., from the moment of penetration of the ovum by the spermatazoon, thereby placing artificial and intentional abortion in the same category as wilful murder.[75]

Non-Western Buddhist opinion

In his selection of opinions on abortion, Jones restricts himself to the views of Western Buddhists. Earlier in the book, however, he describes as 'among the best guidelines we have' a fourteen-point moral code taken from the Vietnamese Tiep Hien Order.[76] The most renowned member of this school is Thich Nhat Hanh, another exponent of 'engaged' Buddhism. If it is legitimate to quote the opinions of contemporary *non*-Western Buddhists on moral issues (and one cannot see why it should not be) we might also cite the following two opinions which were both given by non-Western Buddhists in response to questions from Westerners on the morality of abortion. The first is the opinion of the Dalai Lama as expressed at Harvard in 1988:

[Q] How do Buddhists feel about abortion?
[A] Abortion is considered an ill deed of killing a living being. With respect to monks and nuns, there are four types of ill deeds that bring about a defeat of the vow itself; one of them is to kill a human being or something forming as a human being.[77]

Also relevant is the following exchange between Lama Lodo and a questioner which took place in San Francisco in 1978:

[Q] Before coming to the teaching this evening I spent some time with a friend who is pregnant and has decided that if she doesn't have a miscarriage this week, she will have an abortion. I was wondering if you could say anything about any special way of being of assistance to both the baby and the mother.
[A] The best thing for you to do would be to try to talk her out of the abortion because it is an act of profound negative consequences to kill a human being. A human being's body is so precious that it would be better if you talk her into having the baby and then putting it up for adoption.[78]

In the light of the above the views quoted by Ken Jones cannot be taken as representative of Buddhist opinion, Western or otherwise. Furthermore, it should be pointed out that Jones omits to quote part of the Shin document which expresses a more traditional view on abortion: 'abortion, the taking of a human life, is fundamentally wrong and must be rejected by Buddhists.'[79] It adds: 'The life of the fetus is precious and must be protected.'[80] It should also be noted that the final paragraph of the Shin statement, already quoted above, condones neither situational morality nor abortion. What it acknowledges is the fact that it is the woman carrying the fetus who faces the moral choice. It does not affirm either that the woman would be morally justified in choosing abortion or that she is the only one who can determine whether abortion is morally right or not in her situation. In the light of this it would seem that the Shin statement should not be read as endorsing a situational approach to the morality of abortion, and we therefore exclude it from the remainder of the discussion.

Moral agnosticism

Turning to the views of Roshis Kapleau and Aitken, we note that neither makes a definite judgement on the morality of abortion. Roshi Kapleau has elsewhere written that 'abortion is a grave matter', but adding 'There is no absolute right or wrong, no clear-cut solution.'[81] This is in line with the view expressed by Jones that Buddhist morality is 'situational'. If so, it follows that Buddhism neither has, nor can ever have, a definite position on abortion, or for that matter on any other moral issue. However, there is little evidence that mainstream Buddhism has ever adopted such a *laissez-faire* approach to moral issues. The Buddha himself seems to have held that an objective appraisal of moral choices is possible. At one point he tells Ananda that he is not a teacher who (to use a modern idiom) 'tiptoes around the issues', but one who actively urges and restrains his followers where necessary.[82] Elsewhere it is made clear that the aim of Buddhism is to give clear guidance so that a person can put doubt behind them (*tinnavicikiccha*) and know for sure what is morally right (*akathamkathi kusalesu dhammesu*).[83] These statements presuppose objective moral standards, which situationalism denies. Exponents of situationalism have yet to set out the principles upon which their understanding of Buddhist ethics is based. We might tentatively suggest, however, that the situational approach is a hybrid which combines the Mahayana emphasis on compassion with a Zen-derived emphasis on personal intuition Although little research has been done into Zen ethics it woula appear that its depiction as 'situational' depends to some extent on a view of Zen itself as antinomian and 'bibliophobic' which appears to exaggerate certain aspects of the tradition.[84]

Situationalism and abortion

The view of abortion expressed by Ken Jones, and to some extent reflected in the two above opinions of the Zen masters, differs from that found in the early canonical sources in a number of ways. Jones describes abortion as a 'moral dilemma', but this is not how it was portrayed in the sources we examined. There, abortion was not portrayed as a dilemma but as *an act which should never be done*. The sources are not in the least ambivalent in this respect. It is therefore not correct to describe abortion as a moral dilemma for *Buddhism*. This is not to deny, of course, that the issue may be personally troubling or problematic for many Buddhists.

Morality and meditation

A second point of difference concerns the process by which ethical decisions are to be reached. Roshi Kapleau emphasises the role of *zazen* in determining the right course of action. This is a meditative state in which truth is intuitively apprehended; thus in a moral context one 'sees' what is right. While most schools of Buddhism would not wish to deny the validity of this experience they would also maintain that the proposed course of action should be justifiable independently of it. Thus while intuition might provide a shortcut to the right answer, the conclusion reached should be in harmony with the four criteria mentioned in Chapter 1. It is noteworthy that neither the Buddha nor Buddhaghosa mentions meditation when setting out the steps to be followed in testing the correctness of an opinion, and both insist instead on the primacy of scriptural authority. Reasoning, reflection and intuition (*anubuddhi*) have a place, but as noted earlier are regarded as the *weakest* of the four grounds for validation; it was strongly recommended that any judgements arrived at in this way should be carefully checked against all three higher authorities. It would thus seem that while intuition has a place in moral judgement, it should by no means be relied on exclusively. The early sources suggest that meditative experience is not self-validating and give examples of wrong conclusions being reached through it. For example, the first of the sixty-two groups of 'wrong views' listed in the earliest locus of heretical opinion, the *Discourse on Brahma's Net*, is explained as due to the misinterpretation of meditative experience (*samadhi*).[85] There are also problems of a more practical kind. For example, what happens if someone meditates but no intuitive solution is forthcoming? Can the practice be relied on by one who is confused and distressed? How are non-meditators (and this includes many Buddhists) ever to make a moral choice? Finally, there is the problem that if abortion can be justified through *zazen* there seems no reason why infanticide and other breaches of the first precept could not be justified in the same way.

Abortion and 'life'

Before leaving the topic of situational morality there are a number of miscellaneous points in Roshi Aitken's statement which call for comment. Aitken refers to an aborted fetus as a 'bit of life' which 'goes with our deepest love'. The imagery here depicts life as a stream, an 'ancient process' in which bits come and go continually in the 'flow of life and death'. Although the language here is clearly intended as poetical rather than philosophical, metaphors of this kind must be treated with circumspection. If the metaphor of water is to be used at all it would be more accurate to say that life comes not in a stream but in individual droplets. Life never exists in the abstract, only in the concrete. It is only ever found in individual living things. To talk about life in the abstract is to confuse ethics with ecology. Ethics is not about the grand questions of life in its cosmic form but the mundane business of moral choices which have an effect on ourselves and other individuals. Individuals live and die as organic wholes, and death is always an event in *someone's* biography. In the context of abortion, therefore, a fetus is more properly described not as a 'bit of life' but as an individual being.

For the same reason, the reference to the woman 'killing a part of herself' stands in need of qualification. The fetus is not a 'part' of the woman in the same way that the other parts of her body, for instance her kidneys or liver, are. The fetus is an ontological individual whereas the woman's organs are not. The correct analysis of pregnancy from a Buddhist perspective is not that the fetus is a 'part' of the mother but that one individual is temporarily housed within the body of another. Abortion is therefore neither simply the loss of part of the mother nor a temporary redirection of the life flow. In the simplest terms it is the intentional destruction of a karmic being.

We may note in conclusion a unique implication in Buddhism as far as support for abortion is concerned. Arguments in favour of it typically focus on the dissimilarities between adults and embryos, and are at pains to minimise the grounds for an identification between them. Few people consider that embryonic life is a condition they will ever return to. The logic of rebirth, however, means that Buddhists will expect to undergo this experience many times. Buddhists who condone abortion, therefore, implicitly consent to it being practised upon themselves. In accepting the principle of the 'woman's right to choose', for example, they also accept the correlative status of victim to which such a right condemns them should they be reconceived as a baby which is unwanted by its mother.

The 'Buddhist view'

The second question we wish to consider concerns some of the broader issues raised by the views on abortion expressed by Ken Jones. This relates to the problem of what may legitimately be advanced as a 'Buddhist view' on ethical issues. We may enquire first of all as to whether the views presented are offered as (i) *the* Buddhist view, (ii) *a* Buddhist view, or (iii) a Western situational hybrid. The views on abortion cited have two features in common: they are expressed by Westerners rather than Buddhists from traditional Buddhist cultures, and there is a clear sectarian bias in favour of one school (Zen). Although Roshis Kapleau and Aitken are clearly writing from the perspective of a single school, no indication is given by Ken Jones that what is being presented is in fact a sectarian view. The implication seems to be that situational morality and pro-choice views on abortion are offered as examples of (i) above.

In the light of the criteria we proposed in Chapter 1, however, these views would fail on all counts as candidates for the Buddhist view. In the first place we are given no canonical authority for the conclusions reached. Indeed, it is doubtful whether a single line of scripture could be produced which directly supports the position advocated. Second, there is no attempt to produce secondary or commentarial evidence in support. Third, there is no evidence that the view is pan-Buddhist (indeed, there is strong evidence to the contrary). Fourth, there is no evidence that the view has a broad cultural base. In fact the cultural base here is extremely narrow, and all the views cited stem from Japanese schools. Fifth, we are given no clear idea for how long or how consistently this view has been held.[86]

Abortion in Japan

In connection with the fourth point there is a need for special
caution when considering attitudes towards abortion which have
been influenced by Japanese culture. Indeed, Japan is an anomaly in
several ways as far as ethics is concerned, as we shall see. The country has a very permissive abortion policy and an extremely high rate
of abortion. Writing in 1988 Bardwell Smith reported that a conservative estimate would put the figure at about one million per year.
Other estimates put the figure at close to a million and a half. This
compares to annual figures of around 1.6 million in the United
States and 170,000 in the United Kingdom. The population of Japan
is around 120 million, roughly half that of the United States and
twice that of the United Kingdom. Bardwell Smith notes: 'It is common for women to have had at least two abortions by the time they
are forty years old.'[87] The complex cultural and social reasons
behind these statistics have just begun to be explored, and we
cannot go into all of them here. The explanation does not seem to lie
in the fact that Japanese women are more 'liberated' – on the
contrary, they seem to be the victims of a complex set of circumstances unique to Japanese society.[88]

The *mizuko*

One of the factors which seems to have played an important role in
shaping Japanese attitudes to abortion is the traditional view of the
fetus. The background to this belief (which owes nothing to
Buddhism) has been explored recently by William LaFleur. LaFleur
cites as a typical Japanese response to the morality of abortion an
opinion similar to that expressed by Philip Kapleau and Robert Aitken.[89] Central to Japanese perspectives on abortion is the concept of
the *mizuko* or 'water child'. This is the name given to a fetus which is
aborted or, in former times, to a child which was killed at birth.
LaFleur explains how traditional Japanese culture conceptualises
the fetus as a being whose existence is fluid and indeterminate. A
fetus or a young child is thought to exist partly in the human world
and partly in the spirit world, at a point where the boundary
between the two is ill-defined and easy to cross. Against this background abortion could be seen as a less serious matter than it might
otherwise be. It can be argued, for example, that it does not involve

the killing of a 'full' human being since the spirit is not yet completely committed to human existence. LaFleur writes:

> The child who has become a mizuko has gone quickly from the warm waters of the womb to another state of liquidity. Life that has remained liquid simply has never become solidified. The term suggests that a newborn, something just in the process of taking on 'form,' can also rather quickly revert to a relatively formless state.[90]

In terms of this belief abortion can be seen not so much as the destruction of life as its postponement, a gentle nudging of the fetus back into the world of the gods (*kami*) from whence it came. 'Since the newborn or the fetus was often referred to as *kami no ko* or a "child of the gods"', writes LaFleur, 'it became possible to see a forced return of that child to the sacred world as something within the realm of moral possibility.'[91] Despite this mitigating factor, however, Japanese Buddhists remain ambivalent about abortion. While individual opinions vary, the concept of the *mizuko* is not seen as providing a complete justification for abortion. Nor do the Japanese resort so readily to euphemisms when discussing abortion. LaFleur writes:

> The Japanese tend to avoid terms like 'unwanted pregnancy' or 'fetal tissue.' That which develops in the uterus is often referred to as a 'child' – even when there are plans to abort it. Many Japanese Buddhists, committed by their religion to refrain from taking life, will nonetheless have an abortion and in doing so refer to the aborted fetus as a child, one that clearly has been alive.[92]

Mizuko kuyo

Domyo Miura has been closely involved with the development of the religious service known as *mizuko kuyo* which is held for aborted children. A Buddhist priest, he is the fifty-sixth *Monzeki* of the ancient Enmanin temple near Kyoto and was the first Chairman of the Japan Buddhist Society. In his book *The Forgotten Child*, Miura relates how the need for a memorial service for aborted children became clear to him in the early 1980s from his experience of counselling large numbers of people whose difficulties seemed related to the problem of an earlier abortion. To coincide with the

1000th year of the foundation of the temple in 1981, the Enmanin launched a campaign including advertisements in newspapers and appearances on radio and television to promote the *mizuko kuyo* service. The service itself was the revival of an ancient rite. The popularity of the service was astonishing, and the social phenomenon of *mizuko kuyo* has been much studied since. What is relevant for our present purposes, however, is that Miura does not see the notion of the *mizuko* as in any way minimising or mitigating the seriousness of abortion. In fact, his views on fetal life are remarkably consistent with those we have encountered in the early sources. He writes:

> In the Buddhist scriptures man's existence between life and death is divided into ten stages: these are divided again into two, five stages inside the womb and five stages outside. To describe these: when life is produced in the womb at conception, this is termed 'maku no toki' (the membrane stage); the second stage is 'awa no toki' (the bubble stage); the third is 'ho no toki' (the blister stage); the fourth is 'nikudan no toki' (the flesh stage); the fifth is 'shi no toki' (the limb stage). These are the five stages inside the womb.[93]

Miura is quite clear, furthermore, that Buddhism rejects the conclusion which some draw from the *mizuko* concept, namely that fetal life lacks full moral status:

> In ancient times ... the Buddhist scriptures were already explaining the process of human formation by means of the five stages within the womb, and were thus viewing this as already being life ... There are some people who do not regard what is inside the womb as life, but call it a *mizugo* only when it has left the womb and been delivered either as an abortion, a miscarriage or a still-birth. This is not so. A *mizugo*'s existence begins at the 'membrane stage' produced by conception, which in modern parlance means that life is seen to start from the instant the sperm joins with the egg.[94]

The moral implications of this are clear:

> Buddhism takes the stand that the right to life for all beings must be respected. For example, even before a child comes into the world, Buddhists regard it as a life from the very instant when consciousness is born. The formation of a child's mind and body is said to begin in the womb, and this is something that cannot be

overlooked. Out of billions of sperms one unites with one ovum, life starts, and, as explained in the Nirvana sutra, normal birth takes place after 266 days.[95]

In the light of the above abortion is not morally different from infanticide:

> Certainly killing one's child and throwing away the body is murder, and, in anyone's view is an unworthy act of a human being; but abortion too, even though one doesn't see it and actively lend one's hand to it, is no different from killing one's own child by one's own will and decision.[96]

It would seem from the above that the views of some Japanese Buddhists at least have not been greatly influenced by indigenous attitudes towards the fetus. No doubt there is much variation, and LaFleur suggests that 'most mainline Buddhists in Japan seem to support legalized abortion'.[97] The point to note for our present purposes, however, is that the notion of the *mizuko* involves a view of fetal life which is culture-specific to Japan, and it would not be surprising to find that Buddhist opinion had been influenced by it to some degree. It is for this reason that we highlight the danger of adopting views which have a narrow cultural base as representative of the Buddhist position. Ken Jones, in offering exclusively Japanese views on abortion as examples of the Buddhist view generalises from an unacceptably narrow sample and arrives at conclusions which are not representative of mainstream opinion.

Other Buddhist cultures

Having considered the situation in Japan, it may be instructive to look at the position on abortion in other Buddhist cultures. In 1969 Trevor Ling published a paper which examined Buddhist influences on population growth and control in Thailand and Sri Lanka, both traditional Theravada countries. Ling detected a more conservative attitude towards abortion in Sri Lanka than in Thailand, but found an overwhelming consensus that abortion as a method of limiting population was morally wrong. 'It is quite clear', he wrote, 'that Buddhists in Ceylon and south-east Asia, both monks and lay people, disapprove strongly of abortion as a means of population control.'[98] Ling reports a study by Robert Burnight which showed that of his 960 Thai respondents (all married women) '95.6% were opposed to abortion simply as a means of limiting family size.'[99] 91.8 per cent were in favour of abortion when necessary to save the mother's life, and 12.7 per cent were in favour of it for unmarried women. Ling sheds some light on this last statistic by noting that 'There is a strong emphasis in Thai Buddhism on the importance of the family context in the rearing of children.'[100] No figures are quoted for Sri Lanka but the disapproval rating there would most likely be higher.

Can there be a 'Buddhist view'?

In the course of his discussion Ling mentions a number of points which might lead us to question the notion that there can be such a thing as a 'Buddhist view' on abortion. In an aside he draws a contrast between the Theravada countries and Mahayana Buddhist Korea, where 'abortion is illegal, but widely practised and socially accepted'.[101] Unfortunately he gives no further details of the situation in Korea so we must exclude it from our discussion. Ling also mentions 'a popular Thai belief' concerning the *khwan*, the name given to the spirit which takes rebirth as the new infant. In terms of this belief 'it is only after three days from birth that the *khwan* or spirit has established itself in the new baby, and that the child is "established" as a human being'.[102] This belief (reminiscent of the Japanese *mizuko*) is described as 'rooted in indigenous, non-Buddhist religion'.[103] A final point mentioned, which seems to drive a wedge between theory and practice, is that when asked if they had ever taken action to restart menstruation after having missed a period, 25 per cent of the Thai women questioned admitted that they had, and some mentioned that they had taken herbal preparations known to have an abortive effect.

Moral *bricolage*

Some commentators see discrepancies of the above kind as evidence for the view that ethical principles are arrived at in a haphazard sort of way. A moral tradition evolves, it is suggested, by taking up new principles and modifying old ones in the light of changing circumstances. The direction in which a tradition evolves is determined largely by local pressures, and there are no moral axioms which cannot be modified or discarded should the need arise. Evidence for this theory is provided by the variation in beliefs which is found both within and across different cultures. LaFleur, for instance, regards the final detail reported by Ling above (that some Thai women admit to having taken action to restart menstruation after having missed a period) as significant in this respect, and appears to regard it as evidence that the Buddhist attitude to abortion is equivocal.[104] He suggests that 'Moral rules ... are always contexts for negotiation.'[105] The range of attitudes towards abortion in Buddhist cultures, he believes, provides evidence in support of this theory.

> The variation all the way from Sri Lanka to Japan is interesting. It reflects not only a difference in the way the tradition is interpreted but also the impact of other pressures – such as population and density – on the reading of a religious tradition. This is a classic case of what can be called moral bricolage and the differences that result from it. The Buddhists of Sri Lanka are probably one extreme and the Japanese the other.[106]

LaFleur attributes this diversity to a process which Jeffrey Stout has called 'moral *bricolage*'.[107] The term *bricolage* refers to jobs of the kind a handyman (*bricoleur*) might do using odd bits and pieces of material which are to hand. According to Stout:

> All great works of creative ethical thought (and some not so great) ... start off by taking stock of problems that need solving and available conceptual resources for solving them. They then proceed by taking apart, putting together, reordering, weighting, weeding out, and filling in.[108]

The metaphor implies that moral reflection is always of an *ad hoc* nature involving adjustments, adaptations, and concessions in the light of historical circumstances. Stout offers Aquinas as an example of a *bricoleur* for his accomplishment in bringing together into a single whole 'a wide assortment of fragments – Platonic, Stoic, Pauline, Jewish, Islamic, Augustinian, and Aristotelian.'[109] This characterisation hardly seems appropriate in the case of Aquinas, who understood himself as standing within a tradition which held immutable moral principles. A more appropriate metaphor might picture Aquinas as a chef in a long culinary tradition which specialises in a particular delicacy. While succeeding generations seek to improve on the recipe, each chef knows that not everything can go into the pot, and that some ingredients have no place at all. A contrast could then be drawn with the *bricoleur*, who, like a harassed cook, throws all his ingredients into the pot, stirs well and hopes for the best.

Diversity or unity?

Are Buddhist cultures characterised by *bricolage*? The diversity of the Buddhist cultures of Asia is certainly impressive, but in fundamental respects the differences are often more apparent than real. It was suggested in Chapter 1 that what unites the various forms of Buddhism is their adherence to a common moral code. This can be seen more clearly once we look beneath the superficial differences. While there is certainly evidence of what might be described as 'attempted *bricolage*', in most cases the tradition has been successful in keeping undesirable ingredients out of the stew. What we find, therefore, is not so much diversity as a surprising degree of consensus. Indeed, the consensus is all the more remarkable in the face of the cultural variety encountered by Buddhism as it spread. LaFleur speaks of 'variation' and 'differences', but we have seen little evidence of any large-scale variation as far as the morality of abortion is concerned.

Let us review the evidence. According to canonical Buddhism, abortion is wrong. According to Buddhists in Sri Lanka, abortion is wrong. According to Buddhists in Thailand, abortion is wrong. According to the Tibetan Buddhist opinion cited, abortion is wrong. In painting the picture with such a broad brush, of course, much individual detail is missed out. There is undoubtedly variation in the opinion of individuals in the countries mentioned, but the disapproval rating for abortion overall must be consistently high. Even in the most extreme case we have considered, that of Japan, there is little reason to think that Buddhists regard abortion as fully justifiable. That Buddhists may resort to abortion is not, of course, inconsistent with their recognising that what they are doing is contrary to the precepts and morally wrong. The increasingly popular rite of *mizuko kuyo* can be seen in large part as an attempt by women who have undergone an abortion to come to terms with the inner conflict arising from a choice which is known to be against the precepts.

In the case of Thailand, the evidence that one in four of the Thai women questioned in the survey admitted to having taken action at some time to restart menstruation may admit of more than one interpretation. Ling reports this finding from Burnight's survey as follows:

> When asked if they had ever induced an abortion only 3 of the 960 replied that they had. But when at the end of the interview the question was asked in another form, namely, whether they had ever taken any action to re-start menstruation after having missed a period, 25% answered that they had done so. When asked about the means used, certain herbal preparations known locally to have an abortive effect were mentioned, as well as various patent medicines. The discrepancy between the 3 who admitted to abortion when asked outright, and the 240 who admitted when asked in a more indirect way, provides useful evidence of attitudes of Buddhist women in Thailand to abortion.[110]

The significance of this evidence is open to question in a number of ways. The 'discrepancy' noted between the answers to the two questions may be due to the fact that the women understood themselves to be addressing two rather different issues. Ling assumes, without evidence or argument, that deliberately inducing abortion and taking action to restart menstruation are morally the same thing. This need not be so, and action to restart menstruation may have nothing to do either in fact or intention with abortion. Menstruation may cease or become irregular for many reasons other than pregnancy: this is the condition known as amenorrhoea. The intention of some of the women may have been nothing more than the regulation of an erratic menstrual cycle. That some of the herbal preparations used (but not the patent medicines?) were thought to have an abortive effect is of no direct relevance unless it can be shown that the women intended them to have this effect, rather than some other. A bald statement that herbal preparations and patent medicines 'were mentioned', with no further statistics, is an inadequate ground from which to generalise about attitudes overall. For the statistics to be of any value it would be essential to know what particular 'action' had been taken in each case by what proportion of the women, and with what intent. To interpret the responses to the second question as somehow disclosing the 'true' attitude to abortion amounts, in the absence of this information, to little more than second-guessing. There is, moreover, little need to guess at the attitude of these respondents towards abortion since it is revealed by their clearly-stated and almost unanimous (95.6 per cent) disapproval.

The above does not commit us to the view that none of the women questioned had an intent to abort, and it is almost certain that some of them did. Whatever 'discrepancy' remains, however, may be explained more satisfactorily by drawing a distinction between what Buddhists do, and what they believe to be right. It should not be assumed that the two coincide. The women who had resorted to abortion, either by directly inducing one or through restarting menstruation in order to bring one about, could no doubt offer reasons why they had resorted to such action. Perhaps they would mention family, social or economic pressures, and possibly make reference to the *khwan*. What they would be doing, however, was not offering an argument that abortion was morally right, and in so doing contradicting their previous testimony, so much as advancing a plea in mitigation as to why they had failed to keep the precepts.

Bricolage and rationalisation

It is psychologically difficult for people to do what they know or suspect to be wrong unless they can find at least a partial justification for it. In terms of human psychology this is where rationalisation comes into play. We suggest that an alternative interpretation of the concept of moral *bricolage* would be as rationalisation which takes place at a social or cultural level. This understanding of *bricolage* has the merit of being closer to Lévi-Strauss's original use of the term. Lévi-Strauss coined the term *bricolage* to describe the process through which the primitive mind (*esprit*) creates myths out of the raw material which a culture places at hand. This process is analogous to the manner in which the unconscious produces dreams from its accumulated store of memories and impressions. The function of myths, according to Lévi-Strauss, is to mediate conflicts through the symbolic reconciliation of opposing aspects of reality. In other words, myths provide creative solutions to troubling dilemmas. Moral *bricolage*, we suggest, performs the same sort of function, and is essentially an attempt to mediate conflicting moral and personal imperatives. The outcome of rationalisation on an individual level takes the form of a justification for doing something which one knows should not be done. The result is a creative fudge, very much like the workmanship of the *bricoleur*. *Bricolage* comes into play, however, not to challenge moral norms so much as to find justifications for failing to honour them. In fact it reaffirms the validity of norms by pleading for a dispensation from them due to extenuating circumstances. When the same moral dilemmas (such as an unwanted pregnancy) arise again and again, a stock cultural solution may emerge in the course of time. When a culture places the necessary material at hand – such as the concepts of the *mizuko* in Japan, the *khwan* in Thailand, or the notion that the seriousness of an offence varies with the size of the victim – then *bricolage* will see to it that they are usefully employed. The existence of *bricolage*, however, does not show there are no moral norms any more than creative accounting shows there is no company law.

Conclusion

In contrast to LaFleur, we see little evidence of 'a difference in the way the tradition is interpreted' across the range of Buddhist cultures. On the contrary, we find that the tradition has been interpreted with remarkable consistency. The only exception we have seen to this is Japan, and Japan is very much a special case. The differences among Buddhists in Japan over abortion can be traced to cultural influences which owe nothing to Buddhism. Indeed, the question is often raised as to whether Japanese Buddhism is authentic Buddhism at all. LaFleur notes:

> This judgement that Japanese Buddhism is inauthentic, we should note, is quite often made both inside and outside of Japan, by both non-Buddhists and Buddhists alike. It is tantamount to saying that Japanese 'Buddhism' is really a thin veneer over a mind-set or religious view that is, in fact, non-Buddhist, perhaps even anti-Buddhist.[111]

In the course of the discussion above we set out the views of Domyo Miura, which seem remarkably close to those of Indian Buddhism and would certainly qualify as an example of the 'Buddhist view' on abortion in the sense we have defined it. The fact that opinions of this kind continue to be held shows that there are alternatives to *bricolage*.

In the study of particular Buddhist cultures, the testimony of respondents may be of interest from a sociological or anthropological perspective. Opinion polls and questionnaires, however, are not part of the discipline of ethics. To illustrate this it may be helpful to expand our earlier distinction into a threefold one, and recognise the differences between (a) what individual Buddhists do; (b) what individual Buddhists believe to be right; and (c) what Buddhism teaches (the 'Buddhist view'). Ideally, the three should coincide, but more often than not there will be a discrepancy. Buddhists do not always act in accordance with their moral beliefs, and the 'Buddhist view' is not necessarily what a majority of Buddhists in any time or place believe to be right. It is possible for individuals to be wrong, it is possible for communities to be wrong, and it is possible for entire cultures and civilisations to be wrong. The Buddha realised this and felt that even in his own time standards were in decline.[112] He predicted a further decline to the point where the concept of goodness and even the word for 'good' (*kusala*) itself would disappear.[113] Buddhaghosa elaborates on this apocalyptic scenario describing how norms will be inverted, respect will decline, promiscuity will be rife and evildoers praised.[114]

The traditional Buddhist view is that the true Dharma will gradually be eclipsed in the world by a counterfeit. The fact that a counterfeit had become the norm, however, would not alter the fact of it being a counterfeit regardless of how many people believed it to be genuine. This is one reason why Buddhism insists that personal opinion should not be the first resource for deriving ethical principles and emphasises the role of extrapersonal criteria such as scripture. The opinions and practices of Buddhists can be measured against the principles found in scripture,[115] but it must be remembered which is the yardstick and which the thing being measured. Our purpose in considering data on abortion from different Buddhist cultures was not to derive moral principles; instead, it was to examine the empirical evidence for the twin claims that the Buddhist moral tradition has evolved through *bricolage*, and that opinion on abortion in Buddhist Asia is characterised to a significant degree by divergence and variation. We conclude there is little evidence for either.

V   EMBRYO RESEARCH

Reference was made above to the uncertainty of the scientific data regarding the development of the early embryo. This has led to calls by scientific and other lobbies for the legalisation of research on embryos in order to fill the gaps in present-day knowledge of early embryology. What sort of research is envisaged and for what reasons?

Forms of experimentation

Many different reasons are offered why experimentation on embryos should be allowed. Perhaps the most basic is simple scientific curiosity about the process of embryonic development. Those who wish to engage in research, however, usually stress the applied aspects of such knowledge rather than its purely theoretical value. Among the reasons cited are the development of more efficient contraceptives, the alleviation of genetic abnormalities, the treatment of infertility, and the cure of hereditary and other diseases. Although many of these aims would be shared by Buddhism, there is cause for concern about the nature of the research carried out on the grounds that it involves the destruction of embryos. In some experiments hundreds of human embryos have been created and then destroyed by experimentation. Some are used for testing drugs and toxic substances. Other experiments have a Frankenstein-like quality to them. Human embryos have been placed in rabbits, monkeys and sheep, and human–animal hybrids have been created in the laboratory. Among the goals of some researchers are the harvesting of human organs for transplantation and the creation of new species through genetic manipulation.

The results

The results achieved to date through experimentation on embryos provide little reason to think that many of the beneficial goals mentioned above will be realised. The improved contraceptives promised seem likely to be more efficient abortifacients rather than drugs which will prevent conception. Research aimed at the prevention of genetic abnormalities does not presently offer a treatment or cure but only detection, with the subsequent destruction or non-implantation of the embryo which is thought to be defective. There is concern that as specific genes are isolated through the Human Genome Project, there will be pressure to eliminate through early abortion embryos which seem 'undesirable' for one reason or another. Grounds might include physical or mental disability, gender, height, colour of eyes or skin, or homosexuality, which has recently been suggested to have a genetic basis.

In the past twenty years there have been few breakthroughs which can be attributed directly to research on embryos, and a number of experts feel that little can be learned through experimentation that cannot be learned in other ways. There is no evidence that embryo research is a necessary adjunct of *in vitro* fertilisation, and a number of IVF programmes across the world operate entirely without it.[116] As part of its campaign in favour of embryo research, the journal *Nature* called in 1985 for examples of suitable experiments on embryos accompanied by explanations as to why only human rather than other mammalian embryos could be used. Not a single example was published.[117]

The Buddhist position

In Buddhist terms, destructive experimentation on embryos represents a direct assault on the basic good of life and a breach of the first precept. If the goal of the research is theoretical knowledge, it would amount to the subordination of life to knowledge, and as with any instrumentalisation of a basic good would be impermissible. Might not research be justifiable if motivated by compassion for human suffering and a desire to eliminate disease? The notion that compassion for one being can justify causing the death of another is an instance of selective rather than universal compassion. David Stott observes:

> Such a line of argument is clearly in contradiction to *Buddhadharma*. Since when has Buddha advocated the killing of one being for the benefit of another? One might as well argue that one should kill a rich man to make oneself or others happier. Such a 'compassion' is, of course, not the limitless compassion without partiality that Lord Buddha teaches.[118]

Even if the goal was to relieve the suffering of others, then, experimentation would be inadmissible since it would be achieved at the cost of subordinating life to friendship. The point should be made that the good of life is not an impersonal thing but is grounded in the intrinsic bodily good of individuals. This means that 'life' cannot be considered in the abstract in the way that a utilitarian might conceive of a good to be 'maximised'. Thus it would be wrong to suppose, in the context of experimentation, that through the sacrifice of one life as part of a research programme which led to the subsequent preservation of many lives through its beneficial results, the good of life had been respected.

What if the research were for the benefit of the subject itself, for example as a 'therapeutic' experiment? We can imagine a case where an abnormality is detected in an early embryo for which a cure may be available, but in order for the treatment to proceed certain tests must be carried out. Clearly, there is no way an embryo can give informed consent to the procedures in the way a competent adult patient could. The situation here may be compared to that of other patients who for one reason or another are unable to give informed consent, for example young children and the mentally disabled. The conditions to be satisfied in these cases are typically that: the procedure is therapeutic (for the benefit of the patient), the risks and benefits have been carefully weighed, and informed consent has been obtained from the parent or legal guardian. The likelihood of researchers applying these criteria to embryos, however, seems remote.

Research on unanimated embryos

We suggested above that Buddhism is not committed to the view that all embryos are animated. It might be argued, therefore, that experimentation on embryos which have *not* been animated would be consistent with Buddhist principles. There are two objections which might be made here, one of a practical nature and the other theoretical. The practical problem is that there appears to be no way of telling which embryos are animated and which are not. The doubt as to whether an embryo is animated or not does not minimise the seriousness of research upon it: to be prepared to take the chance that an embryo is not animated is, surely, morally the same thing as being prepared to experiment on it even if it is. There is no way of establishing with certainty that an embryo is not animated, and the more developed the embryo becomes the greater the probability that it is. The theoretical objection is that even if it could be established with certainty that an embryo is not animated, it would still be biologically human. If an embryo is not *now* animated, it may have been so at an earlier time, and as a corpse should be treated with respect and not used as an object of scientific curiosity. For the above reasons, the distinction between animated an unanimated embryos makes no practical difference to the Buddhist position on embryo research. In conclusion, it looks as if the principles outlined above and the conditions which Buddhism would wish to impose

even on research for the benefit of the embryo itself would in
practice rule out almost all research being carried out today.

VI   FERTILITY CONTROL

Fertility control may be thought of as having two facets: one
positive, the other negative. In its positive form, fertility control
has the aim of producing a pregnancy, and in its negative form the
aim of preventing one. We discuss both of these facets in turn,
beginning with negative fertility control, which is by far the most
common.

Negative fertility control

We have considered above the Buddhist understanding of conception
and the development of the early embryo. Very little is said in the
early sources, however, about deliberate attempts to prevent
conception. Perhaps this is not surprising, since it is only in very
recent times that efficient contraceptive techniques have been developed and become widely accepted. Whatever the form of contraception used it will involve some means of frustrating the natural process
which would otherwise occur when the three conditions required for
conception are fulfilled. As a starting point it may be helpful to distinguish the two main types of contraception in use today.

(i) *Contraceptive methods:* the most important of these is the 'pill'.
This acts by preventing ovulation, thereby frustrating the second of
the three conditions laid down by the Buddha for conception,
namely that it should be the woman's fertile period. Thus although
there may be intercourse, and the spirit of a deceased person may be
available for rebirth, there can be no conception since there is no
ovum which can be fertilised. Other types of contraceptive include
barrier devices which keep the sperm and ovum apart, such as the
sheath or the cap. Surgical techniques such as sterilisation and
vasectomy are also used, and a contraceptive pill for men may be
available within a few years. Mention is made in the Monastic Rule
of a primitive device known as a *jatumatthaka*. This is said to have
been used by a woman who in lay life was the consort of a king,
before she became a nun. The device functioned as a vaginal insert,
and according to the commentary, seems to have been rodlike in
form and made from a substance possessing medicinal properties.[119]

No details are given concerning its precise mode of operation, but it may have been thought to function as a barrier device.

(ii) Introceptive methods: these techniques differ from contraceptive methods by operating after fertilisation. Devices such as the 'intra-uterine device' (IUD or 'coil') operate primarily, it is thought, not by preventing conception but by impeding implantation of the fertilised ovum in the lining of the womb. It should be noted that many 'contraceptive' pills also act in this way. Pills with a low oestrogen content may suppress ovulation, but if they fail to do so they will prevent implantation. There is thus the risk that this kind of pill will operate in practice as an abortifacient rather than a contraceptive. Some contraceptive pills are intended specifically to work in this way, for example the 'morning after' pill. As such, they are more correctly described as abortifacient, and the ethical questions associated with their use are similar to those discussed above in connection with abortion. Our discussion here will therefore concern the morality of those methods listed in (i) above which are non-abortifacient.

A view not uncommon among Buddhists is that while the use of introceptive methods is wrong, the use of contraceptive methods is morally acceptable. Thus David Stott writes 'Since contraception, as the name implies, prevents "conception" taking place, it is the opinion of my gurus that contraception is permissible. Nevertheless, one must be very clear that the method chosen really is contraceptive.'[120] The grounds for this conclusion are that while the introceptive methods involve the destruction of a newly-formed being, the contraceptive method does not. With the use of the pill no new life is brought into existence and therefore no being is directly harmed.

Cause and effect

It is not difficult to understand the appeal of the above line of reasoning to Buddhists, since it follows the logical structure of many traditional Buddhist arguments. In Buddhist philosophy everything that exists is said to be a product of causes and conditions, and the time-honoured approach to dealing with undesirable states of affairs is to remove the causes which bring them about. To quote the paradigm case, since our ultimate spiritual problem as human beings, encapsulated by Buddhism as *dukkha* or 'suffering', is brought about by craving and ignorance, to eliminate suffering we need to remove these two causes. If we remove the causes of rebirth, namely craving and ignorance, we can prevent the problem of suffering from arising. The overall strategic response to problem-solving in Buddhism is therefore to eliminate the causes which give rise to the problem in the first place.

The fact that the pill eliminates the causes of conception and prevents the arising of an undesirable state of affairs, namely an unwanted child, gives it a strong appeal to many Buddhists by virtue of its similarity to the dominant pattern of philosophical reasoning found in Buddhist sources. The avoidance of conception, it should be pointed out, does nothing to solve the basic problem of rebirth and suffering in the longer term, since the spirit of the deceased person would simply take rebirth elsewhere. However, the formal similarity in the strategy of tracing problems back to their causes and dealing with them at that point holds a powerful appeal for those familiar with Buddhist teachings. This, plus the fact that contraception has no obvious 'victim', leads many Buddhist to feel that the use of contraception is morally acceptable.

Fertility control in the Monastic Rule

Some reason to doubt whether the above approach is the appropriate one, however, is provided by a brief episode narrated in the Monastic Rule in which monks are convicted of wrongdoing for lending assistance in fertility control. Two contrasting cases are reported: in the first a woman seeks treatment for infertility, and in the second a woman requests medicine to prevent conception.[121] In the first case the patient dies, and the monk is adjudged guilty of a 'misdeed' (*dukkata*), a relatively minor offence in the monastic code. In the second case the contraceptive preparation turns out to be fatal, and the monk involved is also adjudged guilty of a 'misdeed'.

The problem in both these cases is to define the nature of the offence. Unfortunately, the text does not spell out exactly what the monks had done wrong and the commentary passes over both cases without a word. Obviously the accused monks were involved in a treatment which led to the death of the patient, but this is not in itself a cause for censure since patients may die even with the greatest of care. The fact that both women died is not germane to the morality of fertility control; in some of the abortion cases the women die and in some they do not. Nor would it seem that the offence lay in intending the death of the patient, for if so the monks would have been found guilty of 'depriving a human being of life' and expelled. This leaves only two further possibilities: either they had been negligent in some way in their treatment, or the nature of the treatment itself was improper.

Medicine, morals and monks

It is not stated in the main text that the medicine given was overly strong (*khara*) as it is in some other of the cases mentioned, so there is no obvious reason for thinking that the monks had erred in their treatment. This suggests that the treatment was wrong for reasons not of a medical kind. Supporting this conclusion is the position which these two cases occupy in the list of offences under discussion. The two cases of fertility treatment are tagged on to the seven instances of abortion considered earlier. The association with abortion suggests that the treatment was wrong for moral rather than medical reasons. The silence of the commentary, if it betokens anything, suggests that the moral issue is not fundamentally different from that in the cases of abortion.

An alternative interpretation is possible. Perhaps it is not fertility control *itself* which is disapproved of here, but the involvement of *monks* in this field of medicine.[122] In other words it is the close association with women which involvement in fertility treatment entails which is improper. From this it would follow that although *monks* should not administer this treatment it would be perfectly in order for a lay physician to do so. However, if the issue here is one of sexual impropriety, it is curious that we find these two cases listed under the offence of 'depriving a human being of life'. There is an alternative category of offence, the first of the four 'Offences of Defeat', which deals specifically with sexual offences. If sexual impropriety had been the issue, we might expect to find these cases listed there instead, along with the other sexual misdemeanours reported.

In neither of the two cases are the technical details of the means of fertility control specified. From our review of early embryology we can conclude that, subject to the three conditions being fulfilled, conception will always take place after intercourse. It would be helpful to know, therefore, whether the contraceptive medicine was designed (i) to prevent the initial mingling of sperm and menstrual blood or perhaps neutralise their generative power in some way; or (ii) destroy the conceptus once it had come into being. There remains the possibility that this information is not supplied because it is irrelevant to the judgement. In other words, it was thought wrong to suppress life in either way, by intentionally frustrating its sexual generation or by destroying it once produced.

While the destruction or suppression of life can be understood as contrary to the precepts, why should it be wrong to administer a fertility drug? The purpose of this treatment is not to suppress life but to bring it into being. The rationale for the decision in this case is not easy to explain. We can only speculate that interference of any kind in the reproductive process was disapproved of. Perhaps it was felt that the side-effects of drugs and medicines was unpredictable, and that the risk of destroying life was always present when attempting to manipulate natural processes which were poorly understood. It may also have been felt that the proper purpose of medicine in the monastery was not the satisfaction of lay desires, such as that of women to bear children. Given the Buddhist attitude towards procreation, infertility treatment would not be high on the list of medical priorities in the monastery.

Although the evidence is inconclusive, there is reason to suspect that canonical Buddhism may disapprove of fertility control in its negative form. Introceptive methods certainly breach the precepts, and contraceptive methods may be against the spirit of the precepts if not the letter. Positive fertility control would appear to be in harmony with both the letter and the spirit of the precepts, but we find reservations about it nonetheless. It is not easy to identify precise grounds for these reservations, which makes it difficult to raise any serious moral objection to its use.

Sex, love and marriage

At this point it may be helpful to take a wider view of the matter and consider the place of contraception in the overall context of sexual relationships and family life. There is little discussion of the principles underlying these matters in canonical sources, but we might begin with some reflections about the purpose of the sexual act. Buddhism has little to say on this topic, being concerned mainly to dissuade people from engaging in it on the grounds that it arouses desires and passions. We may begin by noting the obvious point that the moral status of the use of contraception depends to a large extent on the intentions and motives of those who use it. In assessing the use of contraception as a means of escaping the consequences of promiscuity, the third precept, which prohibits 'misconduct in things sexual' is relevant. This is universally interpreted to mean, first and foremost, adultery, although this prohibition would normally be extended to other forms of 'improper' sexual behaviour such as promiscuity. Since our subject here is not sexual ethics but biomedical ethics, however, the discussion of contraception will be confined to its use in the context of marriage.

Buddhism and marriage

Before we can go much further in this area we require some understanding of how Buddhism regards the institution of marriage itself, since this is the context in which sexual relations and the raising of children will normally take place. By 'marriage' in the West is normally understood a permanent exclusive partnership between a man and a woman. Such would be the pattern in many Buddhist cultures, although there are exceptions. Another difference is that for Christians marriage is a sacrament, whereas it does not have this status for Buddhists. In fact it is one of the *rites de passage* in which monks traditionally do not participate.[123] When we speak of marriage with reference to Buddhism, then, it will be safer to think of this as the culturally-approved institution for the procreation and rearing of children.

In the absence of any extant Buddhist moral theory on the question of sex within marriage, perhaps we could borrow from the traditional Christian view of the matter as an opening gambit. According to this, the sexual act has two legitimate ends: first, as an expression of love for one's partner, and second as an act of co-operation with God in the transmission of life. The Catholic Church teaches that these two ends of sexual intercourse – unitive and procreative – should not be intentionally separated. A couple who use contraception to render their union infertile, it holds, are setting their will against the integrity of the act. Other Christian denominations adopt the view that although ideally both ends should find expression in the sexual act it is acceptable if only the first is present. The view one takes with respect to the use of contraception will depend on the emphasis placed on retaining the linkage between the two. Of the two ends, there seems to be no reason why Buddhism should not embrace the first without reservation. For those who have chosen married life the sexual act is a perfectly natural and legitimate means of deepening the emotional bond between the partners.

What about the second point, that the sexual act should be open to co-operation with God in the transmission of life? Although this question as it stands would have little meaning for Buddhists (since they do not believe in a supreme being as the source of life) it could nevertheless be rephrased to make it more relevant. For example, if we replaced the word 'God' with 'nature' or 'the natural order', its relevance in a Buddhist context would be clearer. Regardless of the metaphysical backdrop against which we view the act of procreation, one obvious consequence is the generation of new life, and we must consider whether and in what circumstances it is morally justifiable to impede the fruition of the procreative act. In view of the limited guidance on these matters in the sources, and the tentative state of scholarship in the field of Buddhist ethics as a whole, it would be premature to attempt to reach any final conclusions on the matter at this stage. What we hope to do instead is raise a number of issues for consideration and further reflection. Two general lines of argument are suggested below which might be used on opposing sides of a debate. The first suggests that Buddhism would regard the use of contraception as immoral largely on the grounds of its interference with the procreative function of the sexual act.

The suppression of life

It was suggested in Chapter 1 that Buddhism regards life as a basic good. If this is so, it would not seem unreasonable to suggest that one should never act intentionally to frustrate its coming into being. There seems, however, to be some ambivalence on this matter both within canonical Buddhism and in traditional Buddhist societies. On the one hand, we are constantly informed that to be reborn as a human being is a great blessing since it is, amongst other things, the most suitable form of existence from which to attain nirvana. To bring children into the world and provide them with this priceless opportunity would seem, on this view, to be a worthy and desirable goal. According to the Buddha's account of conception, intercourse is the door through which beings come into the world. In the natural course of things this door is open, but the use of contraception forces it shut. Its use therefore deprives beings awaiting rebirth of the opportunity to be reborn. Contraception, accordingly, is an act which does not involve just the couple concerned, but also has an impact on third parties. Although none of the five precepts is infringed when using non-abortifacient methods, contraception does not on the face of it fulfil any of the positive aspects of Buddhist ethics: denying the opportunity of rebirth as a human being is certainly not an act of generosity or compassion, for example. We may note that a very popular Buddhist text, the *Discourse on Benevolence* (*Mettasutta*), urges kindness towards all beings whether born (*bhuta*) or seeking rebirth (*sambhavesi*).[124]

On the other hand, as Ling points out, 'there is no scriptural injunction to Buddhists that they should "multiply and be fruitful."'[125] Procreation is not given a high priority – on the contrary, the tradition constantly affirms that a life of abstinence is the higher life. The Buddha himself renounced family life and became celibate. There seems to be something of a contradiction here, but perhaps the contradiction is more apparent than real. While the life of abstinence and renunciation is the higher, not everyone is capable of meeting its demands in their present stage of development. Fortunately, in Buddhism time is always on one's side, and there may be opportunities in future rebirths to devote oneself totally to the celibate religious life. It cannot be maintained, then, that there is a universal duty to procreate, for if so it would mean the end of the celibate monastic life itself. Nor does a duty to procreate follow from the acknowledgement of life as a basic good, for practical reasonableness does not require that an individual promote *all* of the basic goods in their chosen form of life. There is, accordingly, no reason why individuals should not choose to remain single and celibate. What practical reasonableness demands instead is that none of the basic goods should be deliberately undermined in the form of life we choose. It can be seen that the use of contraception by those who have chosen married life threatens to do exactly that, by interfering in the natural outcome and expression of their sexual activity. Seen in this light, the decision to use contraception becomes a choice against life in that it deliberately frustrates its coming into being.

Responsible parenthood

If we look at the matter from a different perspective, however, arguments in favour of contraception can also be found within the framework of Buddhist values. There is the argument from responsible parenthood which maintains that parents should limit the size of their family in accordance with their ability to provide a secure and loving environment for their offspring. Another point is that responsible citizens should seek to limit the demands they impose upon the resources of society by producing children who will require education, medical treatment, and so forth. In the still wider sphere one might point to a global concern about the strains placed upon world resources as a result of population growth. All of these concerns seem laudable and sincere. However good the consequences of contraception might be thought to be, however, they could not be used in Buddhism to justify an act which is immoral in itself.

Contraception in traditional societies

Turning again to the evidence of traditional Buddhist societies Ling observes that in Thailand and Sri Lanka contraception is widely used and appears to attract no moral censure.[126] At the same time, a number of views both for and against contraception were reported. One Thai Buddhist writer argued that any increase in the human population as a whole must be a good thing on the grounds that it is evidence of a general improvement in the moral state of the universe.[127] The reasoning behind this is that since a human birth requires good karma, it must mean that good karma is on the increase when the birth rate increases. Ling notes that this view was dismissed by other Thais as 'too *recherché* by far'.[128] Other reasons for viewing procreation in a positive light include the suggestion that childbearing is a sign of the Buddha's blessing, and that a male offspring would be a potential member of the Buddhist monkhood.[129] Thai Buddhists were in general more relaxed about contraception. According to 'sophisticated Buddhist informants in Thailand' there was 'absolutely nothing in Buddhist theory against contraception'.[130] Although the number of Thai women using contraception was statistically small (2.5 per cent in 1967) informants did not seem to oppose it on religious grounds. The low figure may therefore be due mainly to problems of availability and distribution. It is interesting that around half of the women were using the IUD and that Buddhist medical workers in Bangkok showed a tolerant attitude towards this. Ling suggests, however, that 'this tolerant attitude is at least as much due to indigenous Thai beliefs as to Buddhist'.[131]

Respondents in Sri Lanka were generally more opposed to contraception than their Thai counterparts. A view reportedly held by some rural Sinhalese Buddhists was: 'It is a great sin to prevent pregnancy. You must allow those to be born who are to be born.'[132] Also reported is the view of 'a Buddhist scholar of international standing' who suggested that 'the prevention of conception was equivalent to depriving a being of life, and amounted to a suppression of life.' In his opinion Buddhism sees a human birth as the outworking of karma, and therefore not something to be interfered with.[133] There was no evidence in either country that the promotion of life in itself was a good thing. 'In general', writes Ling, 'there seems to be no positively held view among Buddhists that an increase in the number of human beings is of itself good or desirable.'[134] This seems a little at odds, however, with his further observation that 'the Buddhist reverence for life has played some part' in the high fertility rates in the Buddhist countries of mainland southeast Asia.[135] We thus seem to be left with two divergent opinions on the morality of contraception. Ling sums up as follows:

> While in Thailand it is commonly asserted that there is nothing in Buddhist doctrine against the practice of contraception, in Ceylon opinion even among the sophisticated is less unanimous. There is thus an ambivalence in the testimony of Buddhists on this matter between these two countries.[136]

The empirical evidence from Thailand and Sri Lanka confirms that the two theoretical positions on contraception sketched out above are in fact held in traditional societies. Some opposition to contraception is not unexpected in the light of our suggestion that Buddhism regards life as a basic good. Indeed, it is otherwise difficult to see what objection there could be to contraception when it involves no breach of the first precept. The most sophisticated objection to contraception mentioned was that which saw it as a 'suppression of life'. Ling also makes the point in his conclusion that 'There is in Buddhist theory a certain potential opposition to the practice of contraception; this is due not to any explicit doctrine regarding the metaphysical aspect of conception but to the general objection to violence to a living being, even a potential being.'[137]

Summary

In the face of these two competing approaches to contraception, perhaps the appropriate Buddhist response is to choose the middle way. It must be said that while the principle of the 'middle way' may or may not have some merit on this occasion, it is not recommended as a general methodological principle for practical ethics. If it were, Buddhists would find themselves telling 'half truths' and being 'half a friend'. Applying it in the present context, however, may not be inappropriate. Thus married couples might be said to act in accordance with Buddhist principles if they remain open to the good of life by providing the opportunity for rebirth to the number of children which local conditions (personal and national resources, cultural patterns, etc.) reasonably allow. Given the lack of interest in fecundity we have noted in classical Buddhism, it is difficult to make out a strong case for abundant progeny. In terms of our earlier discussion regarding the twin aims of the sexual act, this conclusion would imply that while the first of the two aims, the expression of mutual love, should always be open to fulfilment, the second requirement, that the act remain open to the transmission of life, should ideally be present but need not be insisted upon in every act. It might be suggested that a couple who use contraception at certain times are not thereby denying the good of life, but prudently planning for its transmission at the most appropriate time in their marriage.[138]

It should be pointed out that the scope for genuine disagreement on the use of contraception is not evidence of *bricolage* or moral relativity. Unlike the case of abortion, which clearly involves a breach of the first precept, there is no clear guidance in canonical sources which would enable us to state the Buddhist position on contraception with confidence. Our dialogue with the scriptures, supplemented by the testimony of Buddhist opinion, has not yielded a clear-cut solution, and it would be premature to advance more than tentative conclusions at this stage.

Positive fertility control

As noted, positive fertility control refers to situations where the objective is for a woman to become pregnant. Two main techniques are employed to bring this about: artificial insemination and *in vitro* fertilisation (IVF). Artificial insemination involves the introduction of semen into the woman's body by means other than sexual intercourse. If the semen is that of the woman's husband (or partner) the technique is known as AIH (artificial insemination by husband). If the semen is that of a donor the technique is known as AID (artificial insemination by donor). The technique of *in vitro* fertilisation is designed to overcome one of the most common causes of female infertility, namely tubal occlusion.[139] The technique involves the removal of an ovum from the woman and its fertilisation in a laboratory dish by the male sperm. Once fertilised, the egg is replanted (after inspection) directly in the uterus and, if it successfully implants, may grow to maturity in the normal manner. As in the case of artificial insemination, the donor of the gametes can be either the woman's partner or a third party.

It is interesting to note that the production of offspring was one of the skills attributed to physicians who belonged to the ancient Indian *Samana* tradition, of which Buddhist monks formed a part. Megasthenes, a Greek ambassador resident in North India around 300 BC, reported of the *Samanas*: 'they are able to bring about multiple offspring, male offspring and female offspring, through the art of preparing and using drugs.'[140] Zysk suggests that this practice 'does not find a parallel in Pali Buddhist records,'[141] although we noted above a case in the Monastic Rule whereby a monk administered medicine to a woman in order to help her conceive. We also saw that the involvement of monks in this matter was judged to be wrongful, although there was some doubt as to the precise reason. It is noteworthy that other medical practices prohibited for monks include enhancing virility (*vassakamma*) and causing impotence (*vossakamma*).[142] The evidence from the canon on this matter, however, is both sparse and enigmatic, which leaves us largely in uncharted territory. In view of this we propose to refer briefly to the attitude of other faiths towards artificial insemination as a prelude to consideration of the Buddhist position.

Other faiths

As with contraception, any discussion of reproductive issues must take into account the wider context of the meaning and purpose of sexual relationships and the rights and obligations of the parties involved. In this respect some cause for concern about the techniques used to overcome infertility is suggested by the cautious approach adopted by other faiths. The attitudes of the main Christian denominations, along with summaries of the views of Judaism and Islam, have been helpfully set out by Anthony Fisher.[143] There is not, of course, complete unanimity on the matter even within individual denominations, but we can discern a fairly narrow spectrum of views within which certain common objections reappear. The severest strictures are found in the official Catholic position which regards all three techniques as immoral since they separate the unitive and procreative aspects of marital sex. In Fisher's words 'Unlike the laboratory making of the IVF child, life-giving within marital life is properly a part of the mutual physical, emotional, intellectual and spiritual involvement of two persons united "as one" in marriage.'[144] Other denominations do not insist so strongly on the link between the two functions of the sexual act, which opens the way to acceptance of some techniques in principle subject to certain conditions.

AIH

Broadly speaking, the technique of AIH has received cautious endorsement from Anglicans, Protestants, Presbyterians and Unitarians, Baptists, and Lutherans. The conditions attached are commonly that the technique should be used as a last resort by married couples who are without children. Amongst Jews it seems that the orthodox are opposed to it while liberals would allow it in some circumstances. Muslims are not opposed to its use by married couples.[145] Some concern has been expressed by Catholics concerning the manner in which the husband's sperm is obtained, namely through masturbation. The objection is that masturbation is an illicit sexual act in that it has no procreative function. Against this it has been pointed out that masturbation in this context does in fact have an explicit procreative goal in view – in fact this is the very reason it is undertaken. Given that it is performed within the context of marriage with a definite procreative purpose, many would raise no objection to it.[146]

AID

When we turn to AID, however, we find widespread condemnation of the technique by virtually all denominations and faiths. A central objection is that the intrusion of a third party by way of donated gametes breaches the exclusive marriage covenant between the partners and violates the 'one flesh unity of husband and wife'. This objection is summed up in the Catholic instruction *Donum Vitae*:

> Respect for the unity of marriage and for conjugal fidelity demands that the child be conceived in marriage; the bond existing between husband and wife accords the spouses, in an objective and inalienable manner, the exclusive right to become father

and mother solely through each other. Recourse to the gametes of a third person, in order to have sperm or ovum available, constitutes a violation of the reciprocal commitment of the spouses and a grave lack in regard to that essential property of marriage which is its unity.[147]

Another area of concern centres on the interests of the child who would be produced. By virtue of the anonymous nature of most AID programmes, the child would know nothing of its genetic origins and background, a right which is now recognised in relation to adopted children. There are also risks relating to the screening of sperm donors who may have unknown physical or mental hereditary conditions, or indeed, as AID programmes expanded, be genetically related in some way to the parents.

Buddhism and reproductive technology

Two aspects of Buddhist ethics might be thought relevant to positive fertility control. The first is that fertility programmes are aimed at the creation of new life, which would suggest they are in harmony with the basic good of life. When they are successful, additional opportunities for rebirth are presented which would otherwise be denied. The second aspect relates to compassion, since through the care and concern of medical staff couples who would otherwise remain infertile can be helped to become loving parents. It is doubtful whether the traditional Buddhist concern for care of the sick is relevant here, since even if infertility is regarded as a sickness, the techniques which have been mentioned are not a cure.

From a Buddhist perspective, AID would not, like adultery, constitute a breach of the letter of the third precept although it might infringe the spirit of it. It seems likely that Buddhism would share the concern of other faiths regarding the intrusion of a third party into the relationship. The ghost of a stranger could threaten the intimacy and stability of the partnership. The fact that the child would be denied knowledge of its genetic parent would be undesirable, as are the risks of unknown hereditary genetic disorders. If the facts surrounding its conception are withheld from a child the question arises as to whether a breach of the fourth precept against lying might arise, since this precept extends to veracity of all kinds. Finally, the depersonalisation of human reproduction through the construction of anonymous sperm banks is a development Buddhism would be unlikely to welcome in view of its conservative approach to family values.[148]

Turning to AIH, some of these objections disappear. The child will know both of its genetic parents and there is no threat to the relationship between the parents through the interposition of an unknown third party. What about the practice of obtaining sperm through masturbation? In normal circumstances masturbation is prohibited by the third precept. However, as always, we must take note of not just the act but also the intention behind it and its status with respect to the basic goods. If the act is viewed as self-giving intended to promote the basic good of life, and is motivated by love rather than the desire for sexual gratification, it may be felt that in these special circumstances it is not caught by the precept.

IVF

If we follow the traditional analytical 'cause and effect' type of reasoning, there would seem to be no reason for Buddhism to object to the technique of IVF in itself since it merely assists nature in achieving its normal ends. After a minor detour nature is once again back on course, and the chain of normal development will resume. There are, however, aspects of the IVF technique which Buddhism would certainly not approve of. Chief amongst these is the creation of spare embryos by drug-induced superovulation. In order to achieve a higher success rate it is common practice to fertilise multiple eggs and then reimplant up to three. Leaving aside the problem of the multiple pregnancies which sometimes arise from this practice (when the 'excess' fetuses are not deliberately destroyed in the womb), we must consider the fate of the 'spare' embryos which have not been implanted. They are usually either discarded or used for research.

These aspects of the technique are unacceptable to Buddhism under any circumstances for reasons mentioned when discussing embryo research. Essentially this is because they involve either the destruction of a living being (a breach of the first precept) or its use as an object for the benefit of others without its consent. Regardless of any benefits (and surprisingly few have yet been shown) Buddhism could not countenance the use of a human subject as an object of research which is not in its own best interests and to which it has not consented. In this respect Buddhism would align itself with the Nuremberg Code (1946) and the Helsinki (1964) and Tokyo

(1975) Declarations of the World Medical Association which state that: 'in research on man, the interests of science and society should never take precedence over considerations relating to the well-being of the subject.'

Freezing embryos

The practice of freezing embryos which often occurs in IVF programmes is also a cause for concern. Only about half of the embryos which are frozen and thawed survive the process, and many of these become nonviable and are discarded. The prospect of an embryo surviving the process and remaining viable is very low, and a survival rate as low as 8 per cent has been quoted.[149] There are, moreover, serious legal problems concerning the 'ownership' of frozen embryos and there are also concerns about the psychological effects upon a child who may be born many years after either or both of its parents are dead. Couples contemplating resorting to this technique should ask themselves carefully about their reasons for wishing to be parents and consider whether their efforts are genuinely altruistic – i.e. for the sake of their future child – or for the satisfaction of their own desires. A selfish desire would not, from a Buddhist perspective, be an appropriate ground for embarking on the venture. Buddhism imposes no obligation on couples to become parents, and the IVF process involves stress, trauma, expense and disappointment. The majority of IVF patients will never have a child and the success rate in terms of healthy live births is extremely low. The failure rate is over 85 percent, and it is difficult to justify the large wastage involved in producing such few live births.

Conclusion

We might sum up the Buddhist attitude to reproductive technology by saying that the use of donor gametes would not be acceptable, and IVF using the couple's gametes could only be countenanced in the simplest cases where the embryos were immediately implanted. In practice this rules out the use of IVF for Buddhists since there appear to be no IVF centres which currently operate under these conditions. A further problem is that the involvement of a medical team means there can be no guarantee that the wishes of the parents would be observed throughout the procedure. This is because physicians would have professional reasons for insisting on a certain course of action in specific circumstances. For instance, if some of the fertilised ova showed evidence of genetic defects it is highly unlikely that any physician would agree to reimplant them in view of the risk of the parents of a child born with abnormalities, or possibly the child itself, bringing an action in negligence. In view of the grave risk of death or handicap to the child which any form of the IVF technique presents it is doubtful whether the use of even the most basic form of IVF is morally justifiable. The conclusion must be that the nature of the techniques in use at most IVF centres today would not be in accordance with Buddhist ethical principles.

3

At the End of Life

Introduction

The moral issues which arise at the end of life are no less complex than those surrounding its beginning. Nor is the task of analysing these issues made any easier by the tendency in contemporary discussions to run many of them together as, for example, by failing to distinguish clearly between euthanasia and suicide, or euthanasia and the withdrawal of futile medical treatment. We will endeavour as far as possible, therefore, to keep the discussion of the important ethical issues at the end of life as separate from one another as possible. Before turning to the moral issues, however, our first task is to arrive at a definition of death which Buddhism would endorse. We discuss first of all the general problem of defining death, and then move on to consider in Section II how death is defined in the early sources. The conclusions we reach regarding the Buddhist concept of death will equip us to address in Section III the moral issues raised by patients in a 'persistent vegetative state' (PVS). In Sections IV and V we consider euthanasia from the perspective of early and contemporary sources respectively.

I DEFINING DEATH

Old age and death are two aspects of suffering (dukkha) which are constantly referred to in Buddhist sources. Buddhist teachings emphasise the inherent impermanence (anicca) of phenomena; as the early sources put it, 'Whatever has the nature of arising, has the nature of cessation.' All forms of organic life have the nature of arising, since they come into being as integrated wholes at a definite moment in time, namely conception. As such they are compounded entities, and according to Buddhist philosophy it is the nature of all compounded entities eventually to lose cohesion and disintegrate. There is a sense in which death encapsulates all the unsatisfactoriness

139

(*dukkha*) of the human condition since it reveals starkly the impermanence of individual life and also the attendant pain and suffering of old age. In this sense death is the paradigm problem for Buddhism since it is emblematic of all the ills to which karmic life is subject.

In Buddhist mythology death and its attendant woes are represented by the figure of Mara, the Buddhist 'devil' who is often depicted in art as either death or time, symbolically holding the world in his grasp. In the Asian cultures it influenced, Buddhism became closely associated with death and was regarded as authoritative in the rites, rituals and practices associated with the end of life. A comparative study of death rituals in different Buddhist cultures would no doubt yield interesting data on the different customs for the treatment and disposal of the dead. By and large, however, there does not seem to have been any particular problem in deciding who was dead and who was not. Only in recent decades due to advances in medical technology has the problem of defining death arisen.

Defining death

Death has traditionally been determined, at least in the West, by the cessation of respiration and heartbeat. Once these vital functions have ceased the failure of all other organic systems follows almost immediately. This means that when death comes it manifests itself to observers as the complete death of the whole individual. This common-sense view has been challenged in recent decades by the development of machines which can artificially maintain bodily functions. Through the employment of these devices the complete and almost simultaneous cessation of respiration and heartbeat need no longer occur. This situation arises most commonly with patients in intensive care, who may be receiving artificial stimulation from a variety of mechanical sources. In the absence of a total collapse of all bodily systems, which the traditional view of death is predicated upon, the question arises as to precisely which organic functions are the critical ones in determining when death has occurred.

The development of new technology has not so much created a new problem as questioned the assumptions implicit in the traditional criteria. What the technology has made apparent is that the determination of death depends not simply upon organic facts but upon a decision as to which organic facts are to be of significance. In a recent discussion of the question Karen Gervais refers to this as a 'decision of significance'. 'Behind the use of any criterion for declaring death', she writes, 'there lies what I shall call a decision of significance, that is, a decision that there is a certain feature (or cluster of features) whose permanent absence constitutes the death of the person.'[1]

Nature itself does not provide a 'magic moment' when everything ceases. Biological activity, for example, may continue for some time after heart and lung activity has ceased. The traditional criteria of respiration and heartbeat have thus been *selected* from a number of possible indicators for life. What is it that these indicators show? The problem of defining death involves conceptual as well as empirical questions. Before applying the operational tests for death, one needs to be clear about what one is testing *for*. The clinical indicators selected will be chosen on the basis of some conception as to what human death amounts to, and this question will itself be closely related to some understanding of what human *life* is. How one defines death, therefore, is not simply a medical or scientific question. Any account of death assumes some prior conception of what it means to be a living human individual. There is a range of competing approaches to the problem of defining death, and Gervais maps out the three main categories into which they fall:

> There appear to be three chief schools in the debate: those who think that the decision about what constitutes the death of the person is biological in nature, and therefore requires a biological argument in its support; those who consider that we are in the realm of the moral, so that our concern is not so much with what features constitute the death of the person as with the determination of the circumstances under which a person ought to be declared dead; and those who take the problem to be ontological or metaphysical in nature, and hence to require the articulation of an adequate theory of personal identity in its defence.[2]

None of these three approaches is free of difficulty, and the arguments involved in each case are too complex for us to enter into at any length here. Gervais herself argues for the third view. There seems no logical reason why elements of each of the three approaches could not be combined, but if we are to select one of the three I think it most likely that Buddhism would align itself with the first. This is because in terms of the approach we have developed so far, Buddhism sees the human individual as constituted by their organic wholeness rather than by their 'personhood'. Of course, if an alternative view of the Buddhist position on what it means to be a living human being were adopted, one of the alternative approaches to defining death might be more appropriate. In view of our earlier conclusions about the Buddhist conception of human nature, however, we believe the first approach will be the most fruitful in the present context.

The concept of brain death

One of the most important developments in recent years has been the formulation of a criterion for defining death in terms of the cessation of all brain functions. In 1968 an ad-hoc committee of the Harvard Medical School produced a report[3] recommending a definition of death in these terms. This definition has since become widely accepted in medical practice and has been incorporated, sometimes with modifications, into the legal definition of death in many countries. The committee saw itself as responding to a specific problem raised by the development of technology, namely what to do about comatose patients who were apparently kept alive only by machine. Are these patients really 'alive' at all, or is it the case that a machine such as a respirator simply 'obscures our vision of death'?[4] In response to these questions the Harvard committee offered 'irreversible coma' as a new definition of death. By this is meant the complete absence of brain functions at all levels including cerebral, midbrain, brainstem and even upper spinal levels. Death is to be declared when (in the absence of hypothermia or nervous-system suppressant drugs, and artificial stimulation) the patient exhibits: '(1) unreceptivity and unresponsivity; (2) no movements or breathing; and (3) no reflexes'.[5] These three conditions can be *confirmed* by an isoelectric (flat) EEG reading, but this is not a condition in itself. When the three tests are carried out in sequence with a negative result the patient can be declared dead and the respirator which is maintaining ventilation and heartbeat turned off.

Did the Harvard committee introduce a *new* definition of death or simply clarify the traditional one? In terms of the above conditions heart and lung activity are no longer decisive. A patient could be declared dead whose respiratory and circulatory systems were still functioning with mechanical support. Does this mean that the traditional criteria were after all only indicators of brain function, and that death has always been regarded as brain death? Unfortunately the committee did not address any of the philosophical issues surrounding the concept of death and instead confined its attention to producing the operational definition set out above. Although helpfully formulating the tests for brain death no definition of death itself was given, nor any explanation as to why brain death should be identical with human death. As Gervais points out, the committee made a 'decision of significance' but failed to justify it.[6]

The Harvard definition refers to the comprehensive cessation of activity at all levels of the brain. It may come about, however, that only certain parts of the brain are damaged such that there are many types or degrees of brain death. For our present purposes two main kinds can be distinguished: total brain death, as envisaged in the Harvard definition, and neocortical brain death, in which only the higher brain functions of the neocortex are irreversibly lost. The second condition, neocortical death, is commonly associated with patients in a coma or persistent vegetative state (PVS) who can breathe unaided but have suffered permanent loss of consciousness. Philosophers who take the view that what is valuable about human beings is their conscious faculties will tend to argue for a definition of death as neocortical death, since this marks the irreversible loss of awareness and self-consciousness. On such a view PVS patients are already 'dead' (although their heart and lungs continue to function normally and unaided) because of the irreversible loss of their cognitive faculties.

Buddhism and cognitive death

An argument that Buddhism would share the above view has been put forward by Louis van Loon in an article on euthanasia which we will make further reference to in section five. Van Loon argues that Buddhism would reject the Harvard criteria and instead define death as cognitive death. The grounds for this are that 'a patient has ceased to exist as a person' before they are dead on the Harvard criteria. From amongst the various cognitive capacities van Loon singles out volition *(cetana)* as the critical faculty from a Buddhist perspective. The distinctive function of *cetana* is making moral choices, which is why it is synonymous with karma. The 'decision of significance' underlying van Loon's concept of death, therefore, is that the value in human life is to be found in the exercise of moral choice. On these grounds he suggests :'The Buddhist[7] would like to see death defined as the stage where a patient has experienced "volition death" – when he has ceased to exist as a human person, which generally occurs upon the irreversible failure of his cerebrum.'[8] With the loss of the higher brain functions "Life" has then ceased to have any human value; "existence" has lost all its meaning.'[9]

How can it be known when a person has slipped from one state into the other? 'Such "personality death"', writes van Loon, 'takes place when the higher cognitive functions associated with volitional mentality are irretrievably lost or destroyed beyond repair.' He accepts it may not always be easy to define with precision when this has occurred.

It may however, not be possible to define 'Volition Death' as an exact clinical situation. It may involve a number of parameters and may require a period of observation sufficiently long to ensure that an irreversible 'a-cerebral' condition has, in fact, occurred. Indeed, 'Volition Death' may proceed along a patch of 'grey' twilight existence, between the 'white' of life and the 'black' of total biological death. But it should be possible to establish an area where this grey positively shades into black, which would then unmistakingly signify the death of the volitional personality, the 'human' being.[10]

Van Loon does not offer any suggestions as to the clinical indicators by reference to which the line between volition life and volition death could 'positively' and 'unmistakingly' be drawn. The proposed 'period of observation', moreover, would be of little use unless one knew precisely what one was looking for. The perspective van Loon brings to the question is an interesting one and his paper appears to be the first contribution to the debate from a Buddhist perspective. The position he represents Buddhism as holding, however, is incompatible with the account we have provided so far of the Buddhist understanding of what it means to be a human individual. Although van Loon offers no textual evidence in support of the concept of death he attributes to Buddhism, it would be premature to reject his conclusions without examining the textual evidence ourselves. To this task we turn in the following section.

II BUDDHISM AND DEATH

Some canonical passages touch upon the question of what distinguishes a living body from a dead one. The following verse passage describes in stock terms the insubstantiality, impermanence and fragility of the body, and makes a passing reference to three factors which distinguish a living body from a dead one.

> When three things leave the body – vitality *(ayus)*, heat *(usman)* and consciousness *(viññana)* – then it lies forsaken and inanimate *(acetana)*, a thing for others to feed on.[11]

The same three items are mentioned in response to a specific question on death in the *Greater Discourse on the Miscellany*.

> How many things, your Reverence, must be absent from the body before it lies forsaken and cast aside, inanimate *(acetana)* like a piece of wood? Your Reverence, when three things leave this body – vitality, heat and consciousness – then it lies forsaken and cast aside, inanimate like a piece of wood.[12]

We have already seen how the fusion of *viññana* with a biological matrix at conception marks the beginning of an individual life. It is not surprising, then, that the end of that individual life will be marked by the departure of *viññana*. What about the two other terms mentioned in the first passage quoted, namely vitality *(ayus)* and heat *(usman)*? Whatever these things are, since they are listed separately from *viññana*, there is reason to think they belong to the biological side of the equation rather than the psychic one. A little light is shed upon their meaning in a slightly earlier passage from the *Greater Discourse on the Miscellany*. The context is a discussion of the five faculties *(indriya)* of sight, hearing, smell, taste and touch. The question arises as to what these five faculties depend or 'stand' upon *(titthanti)*, in the sense of what provides their physical support or organic foundation. The answer is as follows:

> Your Reverence, what do these five faculties depend upon? These five faculties ... depend upon vitality *(ayus)*. Your Reverence, what does vitality depend upon? Vitality depends upon heat *(usman)*. Your Reverence, what does heat depend upon? Heat depends upon vitality ... Your Reverence, it is just like the case of a burning oil-lamp: the light is seen because of the flame, and the flame is seen because of the light. In the same way, your Reverence, vitality exists in dependence on heat and heat exists in dependence on vitality.[13]

The 'attainment of cessation'

The Greater Discourse on the Miscellany explores a matter which is of interest to our present enquiry. This concerns a profound state of trance known as the 'attainment of cessation'. On entering this state almost all the normal physiological processes which take place in a living person are suspended. This makes the state appear very much like death. Various descriptions of individuals in this state of trance are found in the early sources. Paul Griffiths mentions the story of the monk Mahanaga cited by Buddhaghosa.[14] This venerable monk was so absorbed in trance that he was oblivious to the fact that the meditation hall had caught fire. He remained seated in the midst of the pandemonium until the fire was eventually extinguished by the villagers. Griffiths describes the *physical* condition of a person in such a state as follows:

> An individual in the attainment of cessation is conceived of as being without all but the most basic autonomic physical functions. Respiration has ceased completely, and it is likely ... that heartbeat, blood pressure, body temperature and metabolic levels in general have fallen to a very low level ... The physical condition ... is like ... that of a mammal in the deepest stages of hibernation ... and it is possible for the untrained observer to judge the creature dead rather than in hibernation.[15]

Regarding the *mental* side:

> The practitioner in the attainment of cessation is without mental functions of any kind; the condition is 'mindless' and it is explicitly stated that the ordinary mental functions of sensation, perception and concept-formation do not occur in this condition ... Perhaps the closest analogy in Western psychological parlance to this condition would be some kind of profound cataleptic trance, the kind of condition manifested by some psychotic patients and by long-term coma patients.[16]

Of interest for our present purposes is the question of what distinguishes this profound state of trance from death. Could it be the two factors already mentioned, namely vitality and heat? Griffiths confirms this is so when summing up the main features of the state of cessation: 'It is, in brief, a condition in which no mental events of any kind occur, a condition distinguishable from death only by a certain residual *warmth* and *vitality* in the unconscious practitioner's body'.[17] We quote the relevant passage from *The Greater Discourse on the Miscellany* in full below since it contains additional information which will be of interest to us later.

Your Reverence, what is the difference between a person who is dead, deceased, and a monk in the state of cessation? Your Reverence, in a person who is dead, deceased, the bodily functions have ceased and are at rest, the vocal functions have ceased and are at rest, and the mental functions have ceased and are at rest: *vitality is extinct, heat has gone*, the sense-faculties are broken asunder. In a monk in the state of cessation the bodily functions have ceased and are at rest, the vocal functions have ceased and are at rest, and the mental functions have ceased and are at rest: *vitality is not extinct, heat has not gone*, and the sense-faculties are purified. This, your Reverence, is the difference between a person who is dead, deceased, and a monk who has attained the state of cessation.[18]

Commenting on the above passage Buddhaghosa explains 'bodily functions' as 'inhaling and exhaling' *(assasa-passasa)*. If this particular bodily function has ceased and is at rest in this state but may later resume, it follows that a person could remain alive for an extended time without respiration. This means that any Buddhist tests for death would be unlikely to place great weight upon respiration.

The 'life-faculty'

So far reference has been made to two terms, vitality and heat, as indicators for life. In his commentary on the passage above Buddhaghosa introduces a third term when he glosses 'vitality' *(ayus)* as 'the physical life-faculty' *(rupajivitindriya)*. To illustrate the meaning of this, he uses the image of a fire which has been covered over by ashes. The vital functions of a monk in the state of cessation continue at a subdued level in the way that heat is present in the smouldering embers of the fire. The emergence of a monk from the state of trance is compared to flames which are rekindled by blowing or using bellows.[19] By the 'physical life-faculty', therefore, Buddhaghosa seems to have in mind the biological processes which must continue if life is not to become extinct. In the state of cessation bodily metabolism is greatly reduced, but it must continue nonetheless. If it did not, it could never resume its normal level of activity at a later time.

The 'life-faculty' and taking life

Buddhaghosa makes further reference to the 'life faculty' when defining the precept against taking life.[20] The first of the Five Precepts speaks of 'depriving a living being of life' and Buddhaghosa offers clarification as to what this means:

> In order to make the matter clear we must understand what is meant by the terms *a living being, killing a living being, one who kills a living being,* and *the means of killing a living being.* In everyday language *a living being* means 'a creature' *(satta),* but in terms of Buddhist philosophy we mean the 'life faculty'. What is being said is that in attacking the life faculty the living being is attacked. *Killing a living being* refers to the intent which sets up some means for destroying the life faculty: it is that intention to kill which is *killing a living being,* and *one who kills a living being* is a person who has such an intention.[21]

In the Monastic Rule, the phrase 'should deprive of life' is glossed as 'should cut off the life-faculty *(jivitindriyam upacchindati)'.* Buddhaghosa's explanation is as follows:

> The deprivation of life really means cutting off the life-faculty, so in the word-by-word commentary on the text it is said *should cut off the life-faculty, obstruct it, disrupt the continuity.* The phrase *cut off the life-faculty, obstruct it,* means cutting off or obstructing the causal continuity of the life-faculty, and this meaning is revealed by the words *disrupt the continuity. Disrupt* here means 'disconnect'. The life-faculty itself is twofold: there is the physical life-faculty and the immaterial life-faculty. Of the two, there can be no assault on the immaterial life-faculty, and it is impossible to remove it. The physical life-faculty, however, can be the object of an assault and taken away. But in taking that away, one also takes away [indirectly] the immaterial life-faculty and the two cease together since they exist in mutual dependency.[22]

It would appear then, that the physical and immaterial life-faculties are terms which denote the physical and psychic dimensions of human nature. The two exist in mutual dependency and their permanent separation is death. The immaterial life-faculty cannot be the subject of an assault because it is not a physical thing. Nor, since it is immaterial, can it be directly detected or observed. This suggests that death can only be defined empirically by reference to the physical life-faculty. The physical life-faculty is identified by Buddhaghosa with vitality. Vitality is not identified with any particular organic function such as respiration, heartbeat or the circulation of blood, and seems instead to denote the basic biological processes of life.

Prana

A final term we must consider is *prana*. This has a long and complex history in both Indian philosophy and medicine. The venerable Mettanando, a Thai Buddhist monk qualified in Western medicine, describes the function of *prana* in the context of a brief overview of Buddhist anthropology:

> Human existence consists of two distinct parts: conscious mind and physical body. The two are interconnected, for the mind cannot exist without the support of the physical body, and the physical body cannot be cultivated without mental training. During experiences of supramundane states of consciousness or when the body is close to death, this dichotomy becomes obvious, as is documented in accounts of near-death experiences. The human being is a dynamic system composed of visible as well as invisible elements. In addition to the physical aggregates driven by *prana*, there is the higher life of the soul.[23]

The basic meaning of *prana* is 'breath' and by extension 'life'. As 'breath' it has various shades of meaning, ranging from the gross physical process of respiration to the flow of a subtle energy which was thought to regulate the internal functioning of the body. It is one of the 'humours' recognised by both Buddhist medicine and *Ayurveda*. It regulates respiration, heartbeat, swallowing, digestion, evacuation, menstruation, and many other bodily functions. In this capacity it seems to be closely related to the autonomic system. Mettanando has this to say about *prana*.

The Indian theory of life was based on the concept of 'humors' (*vayo*) paralleling later Medieval theories of medicine in Europe

... The bodily functions were maintained by bodily humors, of which *prana* was one. *Prana* is the 'vital force' of life, located in the heart. Its function was to regulate the other humors responsible for bodily functions, such as breathing and swallowing.

There is clearly an overlap between *prana* and vitality, in that what is regulated by *prana* are the basic biological processes of life. The *Treasury of Metaphysics* comments on the meaning of *prana* as follows:

> By *prana* one understands 'vital breath', a wind on whose existence the body and mind depend. One who commits murder destroys this *prana*, in the same way in which one annihilates a flame or the sound of a bell, that is to say, by impeding it from reproducing itself. Or again, by *prana* should be understood the life faculty (*jivitendriya*): when a man places an obstacle to the birth of a new moment in the life faculty, he destroys it and commits the sin of murder.[24]

Summary

We have now examined four key terms which the early sources associate with the basic functioning of living organisms: vitality (*ayus*), heat (*usman*), the physical life-faculty (*rupajivitindriya*), and breath (*prana*). The first two terms, vitality and heat, were described as interrelated and interdependent. It was said 'heat depends upon vitality', but in what sense? Perhaps heat is an epiphenomenon of vitality in the sense that heat is generated by the basic metabolic processes which take place within a living body. But then in what sense does 'vitality depend upon heat'? This is puzzling, and suggests we may be wrong in thinking of the relationship between vitality and heat as a causal one. The relationship between vitality and heat was illustrated by the image of a flame and its light. It was said that 'the light is seen because of the flame, and the flame is seen because of the light.' This is not a causal relationship since a flame and its light are not really two separate things. Perhaps the interrelationship between the two is to be understood in the sense that both are products of the same underlying process of combustion. The text states that it is from the burning or combustion (*jhayato*) of the oil-lamp that these two interdependent phenomena arise. Perhaps the biological processes of life, then, should be pictured as a form of 'combustion' in which vitality and heat are produced. Just as combustion in the lamp is fuelled by its supply of oil, so the 'combustion' of life will be fuelled by a supply of energy which determines how long the flame of each individual life will burn. According to the Sautrantika school, 'vitality' was a name for the karmic impetus or momentum with which each new life begins. This is compared to the energy with which an arrow is shot from a bow;[25] the arrow will travel forwards only until its predetermined momentum expires.[26] Vitality in this sense is not related to any particular organic function and denotes instead the quotient of karmically-determined energy which drives biological life and determines longevity. The heat and vitality of a living body, therefore, would both be products of the combustion which is biological life, a process fuelled by karmic energy.

Whatever understanding of vitality we arrive at, the important point for our present purposes is that vitality and heat are inseparable. If heat is always present when vitality is present, as it would appear to be, then a test for one will also be a test for the other. The coldness of a corpse in contrast to the warmth of a living body will not have failed to impress itself on Indian observers. The permanent loss of heat from the body would thus seem to be the only empirical criterion offered by the early sources as a means of determining death.

We need not pause to review the meaning of the 'physical life faculty' since Buddhaghosa has told us it is the same thing as vitality. The *Treasury of Metaphysics* also confirms that vitality *(ayus)* and the life-faculty *(jivitendrya)* are the same.[27] We noted earlier that the *Treasury* also regards the life-faculty as equivalent to the last of our four terms, namely *prana*. In the majority of contexts these terms seem to be little more than synonyms for one another, and are often used interchangeably.

Prana and brainstem functions

We noted earlier that *prana* is one of the bodily humours concerned with the regulation of key autonomic functions. The essential function of *prana* seems to be the co-ordination and integration of the basic organic processes which sustain life. In this respect Mettanando sees a close correspondence between the functions of *prana* and those of the brainstem:

> This group of interrelated bodily functions attributed to the *prana* we now recognize as bodily functions maintained by the nuclei of the brainstem.[28]

This identification between *prana* and the functions of the brainstem establishes a connection between the two conceptual worlds of Buddhism and modern medicine. If 'life' in Buddhism is defined by reference to *prana* or one of its synonyms (such as vitality), and if the integrating function of *prana* can be identified with that of the brainstem, then it would seem that the modern test of brainstem death is equivalent to the traditional Buddhist one. When the brainstem is alive, all of the four phenomena mentioned in the ancient sources are present. When the brainstem is dead, there is no vitality, no heat, no life-faculty and no *prana*. In other words, the patient is dead. Brainstem death, therefore, can be seen as a modern substitute for the only empirical test which the ancient sources seem to contemplate, namely the test for bodily heat. This conclusion is endorsed by Mettanando. 'The modern Buddhist doctor', he writes, 'may use the functioning of the brainstem to determine clinical death.'[29]

When speaking of 'death', Mettanando is careful to distinguish between neocortical brain death and brainstem death.

> Thus, from the point of view [of] treatment, death occurs in two stages: (1) the irreversible departure of high-level consciousness and (2) the cessation of the physical function. The first case, the irreversible loss of high-level consciousness, is something we often refer to as 'brain death'. When patients enter a coma ... consciousness has withdrawn inside the physical body ... This withdrawn state of consciousness is invisible to doctors and onlookers, although it remains evident in the involuntary nervous system, including breathing and all the reflexes that can be tested by the usual clinical techniques (i.e., dilation of the pupils). The condition of the patient may be called (cortical) 'brain death,' but all indications show that the brainstem remains intact and functioning.[30]

The condition of neocortical death must be distinguished from the state which the early sources have in mind when they speak of the absence of vitality, heat and the life-faculty. This is death proper, and is to be confirmed, says Mettanando, by the death of the brainstem.

> In the case of physical death – a person in whom consciousness has already disappeared – there is no indication that the brainstem remains functioning. The patient shows no sign of respiration or of response to any of the tests of reflexes (such as pupil constriction) used in normal clinical practice. Significantly, if the

patient is on life-support, such as a respirator, which is doing the job of the lungs although there is no further stimulus to the lungs from the brainstem, the patient can be considered clinically dead. All the reflexes – pupil dilation, swallowing and breathing – need to be tested and found negative before the patient can be declared physically dead. When no brainstem function is present, the artificial respirator no longer gives life support, and we are inflating and deflating the lungs of a corpse, because the *prana* has gone and the consciousness has departed for a new existence.[31]

It must be remembered that heat, brainstem activity, respiration and heartbeat are all only indicators for something. Life is not any one of these things, and is no more reducible to electrical activity in the brain than it is to heat in the body. From our examination of the early sources we have learned how to distinguish between life and death, and found a modern substitute for the ancient test of bodily heat. Our task, however, is still not complete. Now that we have a test we must ask what it is that the test is telling us. To say that the test tells us that the patient is dead is circular. We need to know *why* death should be declared by reference to this criterion rather than some other. Mettanando seems to understand the significance of brainstem death as confirmation that 'the consciousness has departed for a new existence'. While this is certainly one aspect of death, to make it central to the concept of death has certain implications for our understanding of life. It implies, for example, that what is fundamental to our idea of 'life' is the fact that *viññana* is 'in residence'. This suggests in turn that what is most 'essential' to human nature is the 'spirit'. We have rejected this dualism in favour of an understanding of human nature as unitary. Our concept of death, therefore, cannot be one which makes the separation between body and spirit central, although it should not be incompatible with such a notion. Instead, our concept of death must be one which gives due weight to the all the dimensions of human nature experienced in life.

Towards a Buddhist definition of death

Let us recall that to arrive at a satisfactory definition of death we must be clear about three things: (i) our concept of death; (ii) the criterion for defining death; (iii) the conditions or tests which will indicate that the criterion has been fulfilled. We may discuss each point in turn.

(i)   The Buddhist concept of death

One way of approaching this issue is to ask what is lost in death that is present in life. Any number of things could be picked out here, from the physical power of motion to the aesthetic experience of looking at a painting. The kinds of things selected will depend upon the conception one holds as to what it means to be human and alive. As noted above, some philosophers regard the relatively sophisticated intellectual capacities enjoyed by 'persons' as the essential feature of human life. On this view, one is to all intents and purposes 'dead' when these capacities are lost. Being alive in this sense means being alive as a 'person', and is not the same thing as being 'biologically alive', although this is required as a precondition. Being 'dead' in these terms (cognitive death) would mean being unable to exercise the conscious powers proper to 'persons'.

We noted earlier van Loon's suggestion that Buddhism would define death as cognitive death. We also noted that a weakness in his argument was that no textual evidence was produced to support it. From the early textual passages we now have examined, moreover, it seems clear that Buddhism would *not* wish to adopt this approach to determining when death has occurred. In none of the sources we consulted did we find a suggestion that the higher mental faculties are to be used as indicators for life. We certainly found no evidence that life was to be defined by reference to the presence or absence of volition *(cetana)*. Indeed, it was specifically stated that a body cannot be considered as bereft of *cetana* until vitality, heat and *viññana* have left it.[32] We can understand why life is not to be equated with volition from our account of the Buddhist theory of human nature in Chapter 1. Although clearly catastrophic in terms of normal functioning, the failure of the upper brain means only that a bodily organ has been damaged. It represents the loss of a capacity analogous to the loss of sight or hearing through damage to the ear or the eye. For Buddhism, the loss of the higher mental functions is not death. On van Loon's criterion, however, death could be declared while *viññana* was still present in the body. We see that any such suggestion is ruled out by the passages quoted above which specifically mention *viññana* as one of the three things which must leave the body before it can be regarded as a corpse. The criteria supplied by our texts, such as vitality and heat, are clearly of an organic as opposed to an intellectual nature. Death is not depicted as the loss of intellectual functions but the biological end of an organism. This suggests strongly that 'life' for Buddhism means biological life and 'death' means biological death.

Is this to be taken as implying that death is understood as the complete cessation of biological activity? To adopt such a position would be incompatible with the test of brainstem death. This is because the nails and hair of a body can continue to grow for some time even when the brainstem is dead. It also the case, though less well known, that the heart can continue beating for some time, perhaps up to an hour, under the same conditions. Rather than focus on biological activity itself, then, it will be more fruitful to reflect on some of the more general characteristics of death.

One thing we can say with confidence is that death is irreversible. Once death has occurred it is final: life can never be recovered and the organism deteriorates rapidly. This deterioration or 'dis-integration' contrasts with the integrated condition of a living organism, which constantly maintains and renews itself. Symptomatic of this loss is the absence of the characteristics we met with in the texts: death as disintegration is marked by the breakdown of the autonomic functions regulated by *prana* along with the disappearance of heat from the body. Two essential elements in our concept of death must therefore be irreversibility and disintegration. If we put these together we arrive at an understanding of death as the irreversible loss of the integrated organic functioning which a living organism displays.

It is easy to misunderstand the significance of the brain in the determination of death. For Buddhism the brain is the organ of consciousness, but as we have seen, Buddhism does not define death by reference to the loss of consciousness. The significance of brainstem death is not the loss of consciousness but the loss of the brain's capacity to co-ordinate the organic functioning of the body. As well as being the support of consciousness, the brain is also responsible for co-ordinating the various subsystems upon which a complex organism depends. With the loss of this co-ordinating function the organism ceases to be a unified whole and can no longer survive. The test for this condition of disintegration is the death of the brainstem, but it must be remembered that what is being declared under this condition is the death of the human being. It does not follow from the use of this test that a human being is regarded as in any sense *identical* with or reducible to their brain, much less its cognitive functions. We have taken the view throughout that Buddhism views individuals as psychophysical wholes, and our understanding of death must accordingly be as the death of the whole psychophysical organism rather than any one of its parts. Earlier we quoted at length a passage from *The Greater Discourse on the Miscellany* and suggested it would be of some interest to us later. The relevance of this passage is in the pointer it gives to the conclusions we have now reached. In its description of a dead body the text refers to the sense-faculties as 'broken asunder'. The term used here is *bhinna*, which means split apart or separated. In other words, the bodily senses of taste, touch, and smell, etc. have become dis-integrated, and their operation is no longer co-ordinated as it would be in a living, self-regulating organism. It is this lack of integration which characterises death and distinguishes it from life. A stock synonym for death in Buddhist sources is 'the break-up of the body' (*kayassa bheda*). This reinforces the idea of death as the disruption of organic functioning. While it is possible that 'break-up' here refers simply to the physical dismemberment or decomposition of the body, this in itself is nothing more than disintegration at an advanced stage.

(ii) The Buddhist criterion of death

While apparently adopting a different concept of death from Mettanando, we agree with him that Buddhism would accept brainstem death as the criterion of death for a human being.[33] Brainstem death means that the patient has lost irreversibly the capacity for integrated organic functioning. Its occurrence means that the capacity for spontaneous respiration has been irretrievably lost, that heartbeat has ceased (or will shortly do so) and that bodily heat will disappear.

(iii) The conditions for death in Buddhism

What we are concerned with here are the conditions under which death can be declared. Our interest now is in the tests which are to be employed to determine whether brainstem death has occurred. We described the Harvard tests earlier, but have no wish to imply that as tests they are either satisfactory or infallible. Indeed, there has been criticism as to their sufficiency and reliability.[34] For instance, the fact that a patient is unresponsive (the first condition) does not mean that he is not aware; one may be aware of external stimuli without being able to respond to them. The second of the Harvard conditions is that the respirator be turned off for three minutes to see if the patient breathes spontaneously. The brain can survive for longer than this without oxygen, and a period of fifteen minutes or more might be a surer confirmation that respiration will not restart. A final problem is that the unusual circumstances of patients who are dependent upon technological devices may render their physiological responses abnormal in some way and so make the tests unreliable. It is also possible that an EEG machine may produce false or phantom readings due to feedback, static or other forms of electrical interference.

From the Buddhist perspective we have seen that the sources describe a state known as 'cessation', which resembles death in many respects. A person in this state clearly has not lost the capacity for integrated organic functioning since at a later point he will regain consciousness along with all the other vital signs. Nevertheless, there is some doubt as to how such an individual would fare if subjected to the Harvard tests. The first test is for unreceptivity and unresponsivity. Given the example of the monk who sat immobile while the meditation hall burned down, it is almost certain that a person in cessation would fail this test. The second is that there should be no movement or breathing. The texts and commentaries inform us that there are no physical or respiratory functions, which means the second test would also be negative. The third test is that there should be no reflexes. From the descriptions of the state we have, it is hard to imagine a test for reflexes producing any reaction. Finally, as a means of confirmation, an EEG reading can be taken. In the absence of subjects to test, however, it is impossible to be sure what readings would be registered by a subject in cessation, or what interpretation would be placed upon them. According to Byrne and Nilges, EEG readings can be affected by low body temperature:

Hypothermia alone can cause an isoelectric EEG. If the core temperature is below 33°C, the criteria for brain death cannot be applied. Body temperature should be restored to normal before considering brain-related criteria for death.[35]

There is every reason to expect that low body temperature would be a feature of an individual in cessation, which must cast doubt upon the reliability of EEG readings obtained from such a subject. The fact that the Harvard tests may be incapable of distinguishing between the state of cessation and death gives grounds for concern about their reliability. While the state of cessation itself must be rare, there may be conditions similar to it which are not as uncommon.

Conclusion

We would define the Buddhist *concept* of death as follows: *death is the irreversible loss of integrated organic functioning*. A concept of death as the irreversible loss of consciousness (the higher mental faculties) is therefore rejected. It will be noted that a person in the state of cessation could not be dead according to our concept of death since integrated organic functioning continues. As well as a concept of death we also require an appropriate *criterion* for death. We would propose: *the criterion for death is the irreversible loss of the functions of the brainstem*. The third aspect of the enquiry concerned the conditions or tests which determine if the criterion has been fulfilled, i.e. whether or not brainstem death has occurred. We noted grounds for dissatisfaction with the Harvard tests, and would favour the more stringent tests set out by the Conference of Medical Royal Colleges in 1976.[36] These tests do not employ the use of electroencephalography, and are better able to cope with abnormal situations of the kind encountered in the state of cessation. They include, for example, the requirement 'There should be no suspicion that this state is due to ... hypnotics'.[37]

The concept of death as the end of integrated organic functioning and the concept of death as the separation of *viññana* from the body are not incompatible. Indeed, there is every reason to suppose that the organic integration an organism displays is to be explained by reference to *viññana*. In this respect a symmetry can be seen between the end of life and its beginning. Just as death is the loss of integration in an organism so conception is the beginning of the integrated organic functioning which characterises the life of an ontological individual.

We believe the above conclusions in respect of death would be acceptable to mainstream Buddhism. An exception must be noted in the case of Japan where the concept of brainstem death has been widely rejected. Once again, as with abortion, there are distinctive cultural factors involved which have little to do with Buddhism. Becker points out that, 'This rejection comes partly from the Japanese association of brain death criteria with organ transplantation.'[38] He explains the distaste for organ transplantation as arising from Confucian teachings which see the body as a gift and the plundering of organs a sacrilege.

III   THE PERSISTENT VEGETATIVE STATE

The condition of patients in a 'persistent vegetative state' (PVS) has been brought into the public arena by a number of legal cases.

Although this book is about ethics and not law, the two subjects have become closely intertwined in modern life, particularly in the field of medicine. If Buddhism is to play a role in Western society, it will before long be called upon to determine its position with respect to the law. In anticipation of these developments, we have chosen to frame the present discussion in a way which will highlight the legal implications which flow from Buddhist ethical principles. In the course of the discussion we will make reference to a recent English legal case, that of Tony Bland, and in so doing hope to initiate a dialogue between Buddhist ethics and Western law.

The most famous case in this area is that of the American woman Karen Quinlan, who was on a respirator for over ten years while legal battles were fought as to whether the life-support machine should be disconnected. Her doctors argued it should not be, but the court allowed her father to make the decision to disconnect it. However, Karen was able to breathe spontaneously and survived for several years. More recently in England, the case of Tony Bland made legal history when the House of Lords ruled in February 1993 that the food provided to him by tube was a form of treatment and could be withdrawn.[39] It must be recognised that the high cost of maintaining patients in this condition, and the related question of the fair allocation of limited medical resources, have also played a part in bringing the issue into prominence. An important and somewhat worrying feature of these cases is that the courts in both the USA and England have shown themselves willing to be influenced by the judgement of doctors in determining the fate of such patients. The trend which is developing is for courts to accede to the request that the supply of nourishment to these patients be cut off, usually resulting in a heavily sedated death by starvation within a fortnight.

The PVS condition

Before going any further it may be helpful to consider in more detail the condition of a PVS patient. Tony Bland was a seventeen-year-old victim of the Hillsborough football stadium disaster which occurred in April 1989. His condition was summarised in court:

In the course of the disaster which occurred on that day his lungs were crushed and punctured and the supply of oxygen to the brain was interrupted. As a result he suffered catastrophic and irreversible damage to the higher centres of the brain. The condition from which he suffers ... is known as a persistent vegetative state (PVS) ... Its distinguishing characteristics are that the brain stem remains alive and functioning while the cortex of the brain loses its function and activity. Thus the PVS patient continues to breathe unaided and his digestion continues to function. But, although his eyes are open, he cannot see. He cannot hear. Although capable of reflex movement, particularly in response to painful stimuli, the patient is incapable of voluntary movement and can feel no pain. He cannot taste or smell. He cannot speak or communicate in any way. He has no cognitive function and can thus feel no emotion, whether pleasure or distress.[40]

There is a good deal of variation among PVS patients. Some may respond to certain stimuli: Karen Quinlan, for example, is reported to have oscillated between a 'sleeplike' and an 'awakelike' state in which she responded to loud noises and painful stimuli. She also yawned, blinked, grimaced, cried out and made chewing motions although apparently totally unaware of anyone or anything around her.[41] The condition may change over time, and it not unknown for patients to improve markedly after months or even years. It is for this reason the state is labelled 'persistent' rather than 'permanent'. The day-to-day condition of Tony Bland and the treatment he received was described as follows:

Mr Bland lies in bed, his mind vacant, his limbs crooked and taut. He cannot swallow, and so ... is fed by means of a tube, threaded through the nose ... His bowels are evacuated by enema, his bladder is drained by catheter. He has been subject to repeated bouts of infection affecting his urinary tract and chest ... A tracheostomy tube has been inserted and removed. Uro-genitary problems have required surgical intervention ... Without skilled nursing and close medical attention a PVS patient will soon succumb to infection.[42]

Many people instinctively feel that existence in this condition is worse than death. It is suggested not infrequently that such patients would be 'better off dead', that their lives are 'not worth living', and even that they are no longer human beings but 'manicured vegetables'.[43] To some extent reactions of this kind are understandable as an initial response, and the fact that such comments are often heard need not be taken as evidence of a settled conviction that such patients should be killed. Judgements of the above kind, moreover, assume that a distinction is to be made between 'life' on the one hand, and its 'quality' (measured in the form of certain experiences) on the other. Underlying this assumption is a dualism which postulates a dichotomy between the bodily life of an individual and their psychological experiences. Bodily life is seen as the platform or stage upon which experiences of varying quality make their appearance like scenes in a play. What value life has is thus to be found in the quality of the performance at any given time, and if the performance is particularly poor, it may be thought that the best thing to do is get up and leave. Conclusions of this kind follow naturally from the concept of 'personhood' discussed in Chapter 1. Whatever its philosophical merits as an account of human nature, however, it is not one which Buddhism shares. To be human, according to Buddhism, means to exist in the manner described in the doctrine of the five categories. This doctrine speaks of *five* categories, not four or three: bodily life is an intrinsic part of a person's being and not just a condition for the exercise of psychological capacities. Human good is understood by Buddhism as the good of the whole human person. Since life is a good intrinsic to each individual the loss of the higher faculties does not mean that human life ceases to be good. Human existence is embodied existence and no distinction of any moral significance should be drawn between the organic life of an individual and their psychological experiences.

From the Buddhist perspective, the PVS patient is a living human being who has sustained injury to part of their physical organism. Such a patient should not in principle be treated differently to any other patient. Buddhist teachings on the nature of the human person have a specific bearing on the PVS condition. From our discussion of human nature in Chapter 1 we can see that the Buddhist analysis of the PVS condition would be that the damage to the physical organ (the brain) prevents *viññana* from functioning in certain of its modes, primarily that of 'mind-consciousness' *(mano-viññana)* or intellectual activity. Irreversible damage to the neocortex is no more significant from an ethical point of view than irreversible damage to any other sensory organ. Mettanando suggests that in the PVS state the *viññana* of the patient may be active in adjusting itself to the new conditions and preparing for death, and that death will come as the conclusion of this process at the appropriate time. He writes:

As already described, the consciousness has not departed, but occupies an interior dimension. Even though comatose patients are helpless with regards to their physical bodies, the conscious mind, which has withdrawn within, may still be working to mentally prepare the patients for death. This process of mental cultivation can continue for as long as the patient still has *prana* present in the body. The work of the withdrawn consciousness will make the after-life destination as fortunate as possible ... The doctor has the responsibility to insure that the patient can continue to cultivate the mind without interrupting the *prana*, and thus avoid jeopardizing the final phase of the dying process.[44]

This understanding of the condition is clearly at variance with the dualistic view expressed by Lord Keith that in the case of Anthony Bland 'The consciousness which is the essential feature of individual personality has departed for ever.'[45]

Feeding as treatment

The legal issue which is commonly raised where PVS patients are concerned is whether medical staff are under a legal duty to provide treatment and nursing care. In the Bland case, it was argued that feeding was a form of medical treatment, and that as a treatment it should be withdrawn on the grounds that it was futile. This is a conclusion which had already been reached by many courts in the United States. Since this seems to be the issue on which many legal judgements turn, we must consider it from the perspective of Buddhist ethics.

We note first that traditional Indian medicine seems to support the claim that there is a close relationship between medicine and food. In fact the five basic medicines sanctioned in the ancient sources are all foods, namely (1) clarified butter (2) fresh butter (3) oil (4) honey (5) molasses. This group of five constitutes the basic monastic *materia medica*, which was supplemented by 'a more extensive pharmacopoeia of fats, roots, extracts, leaves, fruits, gums or resins, and salts'.[46] It was permissible for monks to take any food as medicine provided it was not consumed primarily for nutrition. If the patient was a layman the distinction between food-as-medicine and food-as-nutrition would lose its restrictive significance, since the laity are not obliged to observe the monastic dietary restrictions. If the same substance can be classified as either food or medicine according to context, then, might it not be argued that the food provided via a feeding tube to a PVS patient is medicine? If so, might it not be legitimate to withdraw the supply of a medicine that is clearly not capable of restoring the patient to health?

The fact that food can be used *as* a treatment, however, does not mean that it *is* a treatment. This is something which can only be determined from the context. The test would be whether in the intention of the physician the substance is administered for medicinal purposes: in other words, is the substance administered to cure a disease? Where PVS patients are concerned, it is difficult to regard the supply of food as constituting medical treatment. If it is a treatment, what is being treated? The food administered by tube to PVS patients is not intended to restore them to health. It is not selected for its medicinal or curative properties relative to the condition. The argument that the removal of the tubes represents the termination of a futile medical treatment cannot really pass muster, since the feeding of PVS patients is never intended as a medical treatment at all. It was suggested by Lord Goff in *Bland* that an analogy can be drawn between tube-feeding and mechanical ventilation, but this analogy is unpersuasive. Whereas a ventilator assists a person to breathe, a tube does not help them to digest. Nor does it assist swallowing, but replaces it. The withdrawal of a ventilator, furthermore, does not prevent the patient breathing spontaneously, whereas the withdrawal of a feeding tube can have no other outcome than the certain death of the patient.

'Futile' treatment

The second point to consider is whether the supply of food, if it is a medical treatment at all, is a futile one. Let us concede for a moment that in spite of the arguments advanced above the supply of nourishment is in fact a medical treatment. To assess whether a course of action is futile it is first necessary to establish the putative objective. The futility or otherwise of a course of action depends upon the likelihood of it achieving the desired goal (that the goal itself might be deemed futile is another issue). The implication of the judgement in *Bland* was that the treatment was intended to restore the patient to health, and that it should be discontinued because it was clearly failing to do this. But this conclusion is odd, since it would be unheard of for modern medicine to attempt to treat the PVS condition by prescribing food. If this is correct, it is difficult to conclude that the 'treatment' was futile. The fact that it was never adopted with a particular aim in view *as* a treatment makes it difficult to conclude it had failed. It could be argued further that if the provision of food *was* intended as a medical treatment it was at least partially successful, since the aims of any course of treatment must include keeping the patient alive, and the treatment achieved at least this much. To withdraw a treatment which is at least partially successful without having a more effective treatment to replace it would seem to be a questionable clinical judgement.

The fact that food is administered to PVS patients via the insertion of a naso-gastric tube lends some credence to the claim that it constitutes medical treatment.[47] The insertion of the tube, however, requires no special medical skill and is best regarded as an alternative means of delivery. There is also a fallacy in the assumption that if tube-feeding *is* part of medical treatment then it *cannot* be part of the everyday non-medical care of the kind which patients would receive from their family or friends. Patients in hospital receive treatment *and* normal care of the kind they would get at home: not everything which medical personnel provide is exclusively medical treatment. Methods of feeding are many and varied (oral, intravenous, naso-gastric) but all have the common aim of providing the patient with nourishment. The means of supply does not in itself transform the nature or purpose of the product, nor does the fact of being dependent on others for its delivery diminish the value of the patient as an individual.

To some extent the legal question of whether or not food is a form of medical treatment has overshadowed the underlying moral issue at stake in the case of PVS patients. Beneath the assertion that the treatment is futile is the judgement that the patient's life is futile. This judgement is incompatible with the principle of the 'sanctity of life' which Western law has traditionally upheld. The issues these cases raise are of a moral and philosophical kind, and cannot be decided without a commitment to a particular view of human nature and human good. The courts seem reluctant to make this commitment, and prefer instead to devolve their responsibility to the medical profession. The judgement in the Bland case allows the medical profession to make its own assessment of the value of the life of PVS patients and to terminate their lives if they are deemed 'not worth living'. What is urgently required, however, is serious reflection on the legal and moral principles at stake rather than the legal ratification of whatever happens to be the current consensus amongst a given group of citizens with medical qualifications.

The intention to kill

The morality of actions for Buddhism can only be assessed when we know something of the intention behind them. What, precisely, was intended by the withdrawal of feeding in the Bland case? The majority of the Law Lords were clear in their judgement that the intention was to cause the death of the patient. Lord Browne-Wilkinson, for example, said:

> Murder consists of causing the death of another with intent to do so. What is proposed in the present case is to adopt a course with the intention of bringing about Anthony Bland's death. As to the elements of intention, or mens rea, in my judgement there can be no real doubt that it is present in this case: the whole purpose of stopping artificial feeding is to bring about the death of Anthony Bland.

Lord Mustill stated that 'the proposed conduct has the aim ... of terminating the life of Anthony Bland by withholding from him the basic necessities of life.'[48] Such an intention seems contrary to the Buddhist precept against taking life. We have already considered one interpretation of the precept in our early discussion of the 'life-faculty', and in another place Buddhaghosa defines it as follows:

> *Taking life* means to kill anything that lives. It refers to the striking and killing of living beings. *Anything that lives*: ordinary people speak here of a 'living being', but more philosophically we speak of 'anything that has the life force'. *Taking life* is then the will to kill anything that one perceives as having life, to act so as to terminate the life-force in it, in so far as the will finds expression in bodily action or in speech.[49]

The majority of the Law Lords accepted in *Bland* that behind the withdrawal of food lay an intention to deprive the patient of life. In terms of Buddhist jurisprudence, are PVS patients really 'alive' in the first place, and do they qualify as 'living beings'? From our earlier discussion of how death should be defined, there can be no doubt that PVS patients are alive. They are clearly not corpses, and are judged to be alive by the current medical standard of brainstem death. They are not dependent on life-support machines and are capable of remaining alive for many years if supplied with nourishment. The courts, like the medical profession, are in no doubt that

PVS patients are alive. In the Bland case, Lord Goff of Chively stated:

> I start with the simple fact that, in law, Anthony is still alive. It is true that his condition is such that it can be described as a living death; but he is nevertheless still alive ... There has been no dispute on this point in the present case, and it is unnecessary for me to consider it further. The evidence is that Anthony's brain stem is still alive and functioning and it follows that, in the present state of medical science, he is still alive and should be regarded so as a matter of law.[50]

There is thus agreement between the traditional Buddhist view, and the contemporary legal and scientific views, that PVS patients are 'living beings'.

The next question is whether or not it can be moral by Buddhist standards intentionally to deprive a PVS patient of life. In *Bland*, the Law Lords held that the withdrawal of food was an omission rather than an act. It was, however, an omission to do something which the doctor was not legally required to do, namely continue with a futile treatment. The Buddhist precept quoted above is likewise framed in terms of acts rather than omissions, in that it speaks of the will to kill 'finding expression' in action. Although a distinction can be made between acts and omission for legal purposes it is generally accepted that for moral purposes it has no validity. In Buddhist jurisprudence, the key ingredient in murder is the intention to cause death: whether the death results from an act or an omission is of little importance. If one person sought the death of another and brought it about by depriving them of food, he would be just as guilty of a breach of the precept against taking life as if he had used a gun or a knife. On these grounds Buddhism would hold that a physician has a duty to act for the well-being of his patient, and that to bring about the patient's death by deliberate omission would be morally as grave as causing his death by a deliberate act. To withhold food from a PVS patient with intent to kill would accordingly be a breach of the First Precept.

Buddhism and PVS

What, then, is the appropriate mode of treatment for PVS patients? We have already suggested that PVS patients are not in essence different from other patients, and it follows that the ethics of their care are not different either. An important point which applies to the care of all patients needs to be emphasised. What is prohibited by Buddhist precepts is the deliberate attempt to destroy life: *it does not follow that there is a duty to go to extreme lengths to preserve life at all costs*. There is no obligation, for example, to connect patients to life-support machines *simply to keep them alive*. Nor is there any requirement to perform surgical operations such as organ transplants on PVS patients for the same reason. While the case of each patient must be considered on its merits, Buddhism would have no objection in principle to doctors discontinuing a treatment which was either futile, or excessively burdensome to the patient in relation to its expected benefits. These principles apply even if the death of the patient is hastened. The objectives of medicine can reasonably be limited to (i) the maintenance and restoration of health (or some approximation of it); and (ii) the relief of suffering. For PVS patients the first is unattainable and the second is irrelevant. These patients cannot be restored to health by medical treatment and it is reasonable to restrict such treatment to *those who are likely to benefit from it*.

In the case of PVS patients who have not been declared dead on the criteria of brainstem death, the provision of food and hydration should be continued. There would, however, be no requirement to treat subsequent complications, for example pneumonia or other infections, by administering antibiotics. While it might be foreseen that an untreated infection would lead to the patient's death it would also be recognised that any course of treatment which is contemplated must be assessed against the background of the prognosis for overall recovery. Rather than embarking on a series of piecemeal treatments, none of which would produce a net improvement in the patient's overall condition, it would often be appropriate to reach the conclusion that the patient was beyond medical help.

There can also be a more positive aspect to these tragic cases which should not be overlooked. This is seen in the opportunity they provide to pursue the good of friendship in the form of the maintenance of human communion. For Buddhism all persons, regardless of their physical condition, are worthy of compassion. Buddhism stresses the need for universal as opposed to selective benevolence, and to exclude PVS patients from this care would be arbitrary and unjust. Even unconscious patients can remain the focus of human emotions and be recipients of compassionate concern. They provide an opportunity for others to exercise goodwill and through benevolent treatment of them to affirm solidarity with them even under the most adverse conditions. Seen in these terms, caring for these patients is not a pointless exercise but an affirmation of the bond of social communion between friends. The alternative, to reject and abandon them, would surely be a denial of the universal compassion upon which Buddhism places such emphasis.

IV EUTHANASIA: EARLY SOURCES

Discussion of the persistent vegetative state leads logically into a consideration of euthanasia, for the PVS condition, as noted, is often cited as one example of the many conditions under which life would be intolerable and no longer 'worth living'. First we offer a definition of euthanasia and identify its principal forms. Following this we explore the canonical evidence from the Monastic Rule.

Forms of euthanasia

The essential ingredient in all forms of euthanasia is intentional killing, and since this is usually contemplated in the context of medical treatment we would define euthanasia as: *the intentional killing of a patient by act or omission as part of his medical care.* As to the forms of euthanasia, a preliminary distinction can be made in respect of its active and passive modes. This distinction relates essentially to the means by which euthanasia is administered. 'Active' euthanasia is the deliberate killing of one person by an act, as for example, by lethal injection. 'Passive' euthanasia is the intentional or deliberate causing of death by an omission, as for example, by not providing food or some other requisite for life. Each of these two modes of euthanasia can take three forms: (i) voluntary, (ii) non-voluntary and (iii) involuntary. 'Voluntary' euthanasia involves the request by a legally competent person that their life should be terminated. 'Non-voluntary' euthanasia is the killing of a non-competent patient. The removal of feeding tubes from comatose patients is an example of non-voluntary euthanasia. 'Involuntary' euthanasia is the intentional killing of a person against his will.

Euthanasia in the Monastic Rule

A number of cases in the Monastic Rule have a bearing on the Buddhist attitude to euthanasia in both direct and indirect ways. All of these cases are found in the part of the Monastic Rule concerned with the offence of 'depriving a human being of life', to which reference has already been made several times. The circumstances which led to the promulgation of this precept are of particular relevance to euthanasia.

The text recounts how the Buddha once instructed the company of monks on the theme of the impure (asubha). Contemplation on the impure is a method used in Buddhism to counteract attachment. The practice is designed as a corrective to the unreflective emotional investment made in worldly things. In practising it one might reflect upon the body as impermanent, a thing subject to decay and corruption, and not a proper object of attachment. Having instructed the monks on this theme, the Buddha retired into seclusion for a fortnight. Unfortunately, the monks became over-zealous in their practice and developed disgust and loathing for their bodies. So intense did this become that many felt death would be preferable to such a repulsive existence. Accordingly, they proceeded to kill themselves, and lent assistance to one another in doing so. They found a willing assistant in form of Migalandika, a 'sham recluse' (samana-kuttaka), who agreed to assist by killing the monks in return for their robes and bowls. Migalandika despatched his victims with a large knife, but halfway through the bloody process suffered a bout of remorse. At this point a devil appeared and whispered reassuringly in his ear that by 'bringing across those who had not yet crossed' he was in fact doing right. In other words, by killing the monks he was saving them from the sufferings of samsara. Reassured by this Migalandika resumed his work and killed a large number of monks, up to sixty on a single day.

The Buddha's response

When the Buddha came out of his fortnight's seclusion he noticed the drop in numbers among the monks and enquired as to the cause. When he learned what had taken place he proclaimed the third of the four most serious monastic offences *(parajika)*. This is the prohibition on taking human life, and was announced as follows:

> Whatever monk should intentionally deprive a human being of life, or should look about to be his knife-bringer, he is also one who is defeated and is no more in communion.[51]

It will be seen that this precept prohibits killing even when the person being killed requests assistance in dying. The phrase 'should look about to be his knife-bringer' is a clear reference to the part played in this episode by Migalandika. Migalandika, it will be noted, was doing little more than acting as the instrument of execution: it was the monks themselves who wished to die, and indeed offered Migalandika their robes and bowls as an inducement for his help. Nevertheless, the role of 'knife-bringer' is specifically singled out for condemnation in the precept. The specific ground for the proclamation of the third *parajika* was thus the practice of voluntary active euthanasia. The monks in this case wished to die because they had apparently made the judgement that their lives were 'not worth living' and that they would be 'better off dead'. Their decision might be thought justifiable on grounds of liberty or autononomy, in the sense that as competent adults it was up to them to dispose of their lives as they saw fit. The Buddha, however, did not agree, and took action to prohibit any recurrence of such an episode. This would seem to make it immoral for Buddhists to have any involvement in euthanasia, either by requesting it or assisting in it. Apart from the monastic precepts, it may be noted that the first of the Five Precepts prohibits both killing and *causing* to kill.[52] It follows that both the person who administers euthanasia and the one who requests it would be in breach of the precepts.

Incitement to death

Various cases in which monks play a direct or indirect part in causing death are found in this section of the Monastic Rule. Since they reveal a consistent attitude towards death across a range of different circumstances, we will summarise them briefly. The first case concerns an incitement to death, and relates how a group of wicked monks became enamoured of the wife of a layman. In order to weaken his attachment to life the monks spoke to the husband of his virtues and the pleasures which would be his reward in heaven:

Layman, you have done what is right, done what is virtuous, gained security from fear. You have not done evil, you have not been cruel, you have not been violent. You have done good and abstained from evil. What need have you of this evil, difficult life? Death would be better for you than life. Hereafter, when you die, when your body is destroyed at death, you will pass to a happy bourn, to a heaven world. There, possessed of and provided with five divine qualities of sensual pleasure, you will amuse yourself.[53]

As a result of hearing this the husband began to eat and drink the wrong kind of food and eventually succumbed to a fatal illness. When the matter was reported to the Buddha he excommunicated the monks and expanded the definition of the third *parajika* to include incitement to death:

Should any monk intentionally deprive a human being of life or look about so as to be his knife-bringer, or eulogise death, or incite [anyone] to death saying 'My good man, what need have you of this evil, difficult life? Death would be better for you than life,' – or who should deliberately and purposefully in various ways eulogise death or incite [anyone] to death: he is also one who is defeated, he is not in communion.[54]

Euthanasia need not involve incitement to death by another party. Nevertheless, there are occasions where pressure can be applied in subtle or even unconscious ways such that a vulnerable person may come to feel that death is something they should request. What is wrong with any incitement to death, as in the above example, is the implicit denial that life is a basic good.

Another case concerns a monk who appealed for the swift execution of a criminal:

At that time a certain monk, having gone to the place of execution, said to the executioner: 'Sir, do not keep him in misery. By one blow deprive him of life.' 'Very well, your Reverence,' said he, and by one blow deprived him of life.[55]

The monk's motive, apparently, was to spare the prisoner the mental distress of having to wait for the appointed time of execution. The prisoner was to have been killed anyway, and the monk's intervention simply brought forward the inevitable outcome. In spite of his desire to spare the prisoner suffering, the monk was nevertheless guilty of a breach of the precept.

The final case we will mention concerns a monk who assists in bringing about the death of an invalid by prescribing a drink which will be fatal for him:

> At one time a certain man whose hands and feet had been cut off was in the paternal home surrounded by relations. A certain monk said to these people, 'Reverend sirs, do you desire his death?' 'Indeed, honoured sir, we do desire it,' they said. 'Then you should make him drink buttermilk,' he said. They made him drink buttermilk and he died.[56]

The reason why the relatives desired the death of the patient is not made clear. The circumstances were that an individual had suffered amputation of the hands and feet. This may have been for medical reasons, but as judicial mutilation was not unknown at the time perhaps also as a punishment for some offence. A person in this condition would be able to do little for themselves and would require constant attention and care, including assistance with feeding. The family expressed the opinion that it would be better if the man died. This may have been because they judged his quality of life to be so poor that he would be 'better off dead'. Perhaps their motive was simply to be free of the burden of providing the care and attention he required. It may even have been a combination of these reasons. We are not told if the patient agreed with the view of his family that he should die. The circumstances suggest this was a case of active euthanasia, although it is not clear whether it was voluntary or not. The monk who gave the advice was excommunicated. A similar verdict was pronounced in the case of a nun who recommended a concoction of 'salted sour gruel' (*lonasuviraka*) as a means of causing the death of another patient in the same condition.[57]

Review of the cases

The cases discussed above do not, for the most part, resemble the circumstances in which euthanasia might be sought today. The principles they embody, however, are absolutely central to the morality of the practice. What the above cases show is a consistent pro-attitude towards life in circumstances where its value may be thought in doubt. In each of the cases considered, any course of action involving an intentional choice against life is deemed wrongful. Thus it is wrong to act as 'knife-bringer'; wrong to emphasise the positive aspects of death and the negative aspects of life; wrong to incite someone to kill another, and wrong to assist others in causing death. While we might wish for more detail in each of these five cases there does seem to be a common theme running through all of them: we might sum this up as the principle that *death itself should never be directly willed either as a means or an end.*

Two cases in the Monastic Rule are particularly relevant to the grounds on which euthanasia is sometimes thought justifiable today, namely autonomy and pain. The first case above shows that euthanasia cannot be justified in Buddhism on grounds of autonomy. Euthanasia for the relief of pain is ruled out by the case discussed in Chapter 1. This concerned the monk who was dying in pain; it will be recalled that motivated by compassion for his suffering, his fellow monks 'spoke favourably to him of death'. The decision in these two cases confirms that euthanasia would not be acceptable on either of these grounds. The remaining cases provide further confirmation of the immorality of the affirmation which lies at the heart of all forms of euthanasia, namely that death is better than life.

V EUTHANASIA : MODERN VIEWS

To conclude our discussion of euthanasia we consider some modern interpretations of the Buddhist position. This section provides a review and commentary on three discussions of the topic – those by Kapleau, Lecso, and van Loon – which between them provide a reasonable sample of the range of views encountered in recent discussions.[58]

Philip Kapleau

Philip Kapleau has written two books which discuss death and dying from a Buddhist perspective.[59] The first is a short anthology of writings from a mainly Zen Buddhist perspective on the themes of death, dying, and rebirth. It does not discuss the ethical aspect of these matters and makes no reference to euthanasia. The second volume, *The Wheel of Life and Death*, explores the same themes in an expanded way. It contains a short discussion of euthanasia[60] in conjunction with suicide, and it is suggested that Buddhism would reject the practice of either. Kapleau observes that 'Buddhism is emphatic in its opposition to suicide',[61] and in respect of euthanasia comments:

> Buddhism holds that because death is not the end, suffering does not cease thereupon, but continues until the karma that created the suffering has played itself out; thus it is pointless to kill oneself – or aid another to do so – in order to escape.[62]

Euthanasia and consent to treatment

We noted earlier that confusion sometimes arises in connection with the various forms of euthanasia. Kapleau makes the following distinction between active and passive euthanasia: 'Some believe there is a moral distinction between active euthanasia (removing a feeding tube from a comatose woman) and passive euthanasia (not resuscitating a terminal patient if he has, say, a heart attack).'[63] These are not the most apposite examples. In terms of the distinctions we made above, the removal of a feeding tube is an example of passive as opposed to active euthanasia. This coincides with the view of the courts that the stopping of feeding is an omission (hence passive) rather than an act. The example of passive euthanasia which is given is even more problematic in that it may not really involve euthanasia at all. When discussing 'Euthanasia and the Law' Kapleau writes:

> As I understand it this latter (passive euthanasia) has become generally acceptable legally, that is, a patient's basic right to make moral choices, in this case to refuse life-sustaining treatment.[64]

Two issues have been confused here which must be kept separate: the first concerns a patient's right to refuse treatment; the second is passive euthanasia. The example given in the extract above relates to the first question, namely the issue of consent to treatment, rather than passive euthanasia. It may be noted in this connection that although it has only recently become legally indisputable, under Anglo-American law a patient has always had the legal right to refuse medical treatment, life-sustaining or otherwise. Indeed, it has always been illegal in both the United Kingdom and the USA to administer any kind of treatment without the patient's consent. The decision not to accept or provide treatment, however, must not be confused with passive euthanasia. The distinction between this and euthanasia is that in the latter the death of the patient is directly willed by the person who administers it. In other words, certain acts are done or omitted with the intention of causing the patient's death. This is the essential ingredient in all forms of euthanasia. A distinction must be drawn between this and, for example, the decision not to resuscitate a patient (or not to provide certain forms of treatment). Decisions such as these need not and will not, normally, be made with the intention of bringing about the patient's death.

Intention and foresight

The decision not to treat a medical condition does not mean that the intention of the physician is to cause the death of the patient. While it may be *foreseen* that the patient will die if the treatment is not provided, it is not normally the physician's *intention* that he should die. In euthanasia, it is always the physician's will that the patient should die. This distinction between intention and foresight is an important one, and is in accordance both with common sense and the law. To illustrate the point with a mundane example: a person may go out in the rain *foreseeing* that he will get wet but without *intending* to get wet. Likewise, in a legal context, murder is defined as intentional killing where death is the aim: it is not enough that death is merely foreseen, even foreseen as certain. The distinction between intention and foresight is often blurred by proponents of euthanasia[65] in order to give the impression that acceptance of the principle that patients should have the right to refuse treatments which are futile or too burdensome entails support for euthanasia.

The role of the physician

Perhaps the above distinction will become clearer if it is remembered that the physician's role is not to preserve life at all costs but to restore the patient to health and well-functioning as best he is able. As is often pointed out, the task of the physician is not to treat a disease but to treat a *patient* with a disease. It is what is best for the patient which must determine the course of treatment if, indeed, treatment should be given at all. It is impossible to lay down hard and fast rules regarding treatment since although the diseases may remain the same the case of each individual who has the disease will be different.

Perhaps an example will help. The treatment of pneumonia in an otherwise healthy adult is a condition which would normally be treated vigorously with the aim of returning the patient to health. A case of pneumonia in an eighty-six-year-old comatose patient who has suffered several strokes and also has Alzheimer's disease, however, would be a very different matter.[66] In this case it would be appropriate to ask whether the treatment would materially improve the patient's life overall or simply be a futile attempt to delay for no good reason an outcome which was inevitable. In the latter case, a decision not to treat the patient would *not* count as passive euthanasia. The question of euthanasia would only arise if there was a deliberate intention on the part of the medical staff to hasten the patient's death, for example by wheeling the bed over to an open window in the hope that the pneumonia would get worse and the patient's death be accelerated.[67]

Examples of this kind need not be restricted to comatose or elderly patients: some patients may feel that the treatment for certain forms of cancer involving chemotherapy and radiation are simply too burdensome to undergo when measured against the statistical likelihood of a cure. In these circumstances it is quite reasonable to choose to enjoy a few years of life without painful treatment as an alternative to subjecting oneself to the side-effects of powerful drugs in the possibly remote hope of prolonging life.[68] In certain cases the decision not to treat will be made by the patient and in other cases by the physician. The fact that the physician decides to withhold treatment, however, does not mean that he intends the patient to die, as would be required for euthanasia, even though he believes there is every likelihood of this being the outcome.

It is worth pointing out that the administering of painkilling drugs to terminally ill patients which may coincidentally hasten their death does not count as euthanasia for similar reasons. What the physician is typically endeavouring to do is enhance through medical treatment the condition of the patient overall, in this case by freeing them from pain. What he is *not* doing is choosing against life by willing their death. In contrast, as an example of euthanasia, we can recall the example from the Monastic Rule cited in Chapter 1. According to the commentary the guilty monks 'made death their aim' (*maranatthika*). These monks willed death as a means to the end of pain, and this is the intention which characterises most forms of euthanasia. In the alternative case, above, the physician wills the enhancement of life through the elimination of pain, while accepting that his efforts may hasten the advent of death. Death, however, is neither intended or chosen either as a means or an end.

Phillip Lecso

Reasons why Buddhism would be opposed to euthanasia have also been suggested by Phillip Lecso. Writing from a Tibetan Buddhist perspective he suggests 'The objections to euthanasia would be on two grounds, that of karma and the mode of death.'[69] We may examine each of these grounds in turn.

Karma

Dr Lecso explains his first point by saying 'In Buddhism a terminal illness is not considered a chance event ... a terminal illness represents the repayment of a karmic debt.'[70] The reasoning here is that since the illness is due to the maturation of karma it is undesirable to try to prevent it taking its course. If it were prevented from taking effect in this life it would only have to be faced again at a future point, perhaps in less advantageous circumstances. Better, therefore, to face up to it now and allow it to exhaust itself in this life. What is not precluded, however, is the relief of pain:

> This non-interference with karma, however, does not exclude the compassionate intervention of relief of physical pain with analgesics (of necessity not leading to lethal doses) or to soothe mental distress with sympathetic listening and counsel. For the terminally ill, Buddhism advocates hospice care, not euthanasia.[71]

While this conclusion is a sound one, the argument which leads to it is questionable for three reasons. In the first place, if it is granted that a terminal illness is a karmic debt and should not be treated, there is the practical problem of determining which illnesses are terminal and which are not. Obviously, there are many cases which have a high prognosis of being terminal, but equally, there will be many cases which although diagnosed as terminal turn out not be so. Perhaps in some cases the treatment might even be successful and convert a terminal case into a non-terminal one. Conversely, certain illnesses which were initially thought to be treatable may turn out in the event to be terminal. The practical problem of determining in advance which cases will be terminal and which not makes it impossible to decide which illnesses to treat and which not to treat.

The second problem concerns the validity of the distinction between therapeutic treatment (which interferes with karma) and the relief of pain (which does not). The question we might ask here is why is the relief of pain not also an interference with karma? If the consequences of the karmic act manifest themselves in a painful terminal illness, then mitigating the pain of the illness would seem to be an interference with the karma at least to some degree.

The third problem is that if there *is* always a causal relation between karma and illness it means that every illness treated is only an illness postponed. If illness is due to evil karma, as this theory suggests, then there is really no significant distinction to be made between terminal and other forms of illness. The argument against the treatment of terminal illness thus becomes an argument for abandoning *all* forms of medical treatment. Why bother to postpone *any* evil karma by ameliorating its effects through treatment of any kind? Why, indeed, bother to build hospitals or practise medicine at all? Since Buddhist monks have long done both we must assume there is not really any fundamental conflict between the treatment of the sick and the doctrine of karma, and this argument must therefore be mistaken in some respect.

The argument against euthanasia on the grounds of interference with karma, therefore, seems to face major problems. The fact that an illness has a karmic cause should have no bearing on the question of its treatment. Speculations about karma cannot be allowed to influence medical treatment. The principle which should govern the decision whether or not to treat any illness is whether it is likely to be of benefit to the patient taking all the known factors (including the burdens of the treatment itself) into account. The physician must be disposed at all times to do whatever is in the patient's best interests, while accepting that in some cases this may mean doing nothing.

At this point, as with our earlier discussion of Philip Kapleau's definition of passive euthanasia, we seem to have moved from the question of euthanasia *per se* to consideration of the ethics of when to treat a patient. We may wonder whether the argument from karma really has any bearing on euthanasia at all. The argument advanced by Dr Lecso seems to involve the following steps: (1) a terminal illness is the repayment of a karmic debt; (2) the repayment of a karmic debt should not be disturbed by the physician; therefore

(3) a doctor should not intervene to keep a terminally ill patient alive. Since the circumstances outlined do not envisage the physician willing the patient's death, it is difficult to see where euthanasia fits into the picture. Unfortunately, no definition of euthanasia is offered by Dr Lecso so it is difficult to be sure how he understands it. We might surmise, however, that he understands it in terms not dissimilar to Philip Kapleau, namely as the decision not to treat terminally ill patients. If so, this may have little to do with euthanasia strictly defined.

The manner of death

The second argument presented by Phillip Lecso is again not really an argument against euthanasia as such, and has more to do with how the terminally ill should be treated and cared for. This issue relates to the use of narcotics as an aid to the relief of pain.

> Next is the argument based on the mode of death. In most discussions as to the mode of euthanasia, especially for the conscious individual, large doses of narcotics are presented as a merciful and idealized way to die. Buddhists would strongly disagree with this ideal of the comatose death. The act of dying and the dying process are felt to be a vital link between this and subsequent existences. The state of consciousness and the level of mindfulness are of crucial importance.[72]

This point is an important one where treatment of the terminally ill is concerned, and is perfectly valid from the point of view of Buddhism. However, its connection with euthanasia is uncertain. The link seems to be that euthanasia is sometimes carried out by administering increasing doses of drugs such as morphine to the point where the dosage becomes lethal. It is certainly true that from a Buddhist point of view it is undesirable for anyone to die with a clouded mind. The argument above, however, is really only against dying with a clouded mind, not against euthanasia as such. Euthanasia can be administered through increasing the dosage of a narcotic, but it can also be performed in other ways. In some of these ways, the patient may maintain perfect mental clarity up to the end. Indeed, one of the arguments advanced in *favour* of voluntary euthanasia is that it allows patients to die before they begin to lose their faculties due to mental incapacity. If euthanasia could be carried out while the patient remains completely mindful, the above argument based on the 'mode of death' would present no objection to it, and might even be used as an argument in support of it.[73]

While both of Lecso's arguments are very much in harmony with Buddhist values, neither of them leads to the conclusion that euthanasia is immoral. The valuable point we understand him as making is that where Buddhism is concerned death must be treated in the context of life. This has two implications. The first is that extraordinary means should not be resorted to in order to preserve life as an end in itself at the cost of losing perspective on its overall meaning and purpose. The second is that since death is the gateway to new life we should aim to approach it in a clear and mindful state rather than in a drugged and comatose condition which essentially represents 'an unconscious attempt to reject death and any meaning that it may hold'.[74]

Louis van Loon

Both Philip Kapleau and Phillip Lecso take the view that Buddhism is opposed to the practice of euthanasia, and emphasise the need to approach death as a meaningful experience. The third view we consider takes a contrary line, and argues that Buddhism favours euthanasia. We have already made reference above to the 'volition death' criterion proposed by van Loon. Implicit in this definition of death is the notion that the value of human life is to be found in the capacity for self-awareness and conscious choice. When this faculty is lost the individual is 'dead'. This is the premise upon which van Loon bases his argument in favour of euthanasia.

Under what conditions is euthanasia justified, and what forms of euthanasia would Buddhism condone? Unfortunately, the answer to these questions is not explicitly stated by van Loon. There is also evidence that he, too, runs together the separate issues of euthanasia, on the one hand, and the withholding and withdrawal of treatment on the other. This is suggested by a comment in the early part of the article:

> Although a Buddhist considers life to be extremely precious, he does not imagine it to be sacred, divine. He is therefore not committed to stubbornly preserving a spent, doomed and suffering-ridden life for its own sake and at all costs. For him there are no 'souls' that can be 'saved' or 'lost' or 'returned' to their Maker.[75]

It is true that while Buddhism values life highly it imposes no obligation upon its followers to 'strive officiously to keep alive', as the famous couplet has it. Nor, for that matter, does Christianity, with which Buddhism is here contrasted. The notion that Christianity is committed to stubbornly preserving life for its own sake and at all costs is a travesty. Both religions are very much aware that 'man that is born of woman hath but a short time to live',[76] and it is because of their recognition that death has a deeper significance in human destiny that they do not insist on clinging to life at all costs. Once again, however, we must ask what this point has to do with euthanasia. As we have noted, it does not follow that not being committed to stubbornly preserving life means acceptance of euthanasia.

Other statements suggest that there is indeed some confusion about what is really at stake as far as euthanasia is concerned. One of these is as follows:

> The debate on euthanasia concerns the ethics of preserving or prolonging life beyond certain limits of human viability; it involves an assessment of the degree of suffering and distress that may be inflicted or should be endured as a consequence of medical treatment that aims to forestall or delay death.[77]

Another example is:

> In short, we should not forcibly extend a suffering-ridden, doomed or volitionless life simply because it is technologically possible to do so, or in the mistaken idea that we have an obligation to 'preserve' the 'divine gift' of such meaningless life for a few more days, weeks, or months.[78]

What van Loon is objecting to here is the imposition of burdensome and futile medical treatment, and Buddhism would have no wish to disagree with him on this point.

Criteria for euthanasia: volition or pain?

Granted that life need not be preserved at all costs, would van Loon still understand Buddhism as being in favour of euthanasia? Apparently so, for he sees many other cases in which it could apply. The list of possible candidates is a large one, and euthanasia 'spans an enormous range of potential applications' from 'unborn children' to 'centenarians'.[79] What makes someone a candidate for euthanasia? In our earlier discussion on defining death we made reference to van Loon's criterion of 'volition death', and it is this criterion which also determines whether a life any longer has human value. It is at this point, however, that the focus becomes a little blurred. Van Loon writes:

> Based on the Buddhist view that volition constitutes a man's essential 'beingness' ... it should be clear that it is from this standpoint that he judges the desirability or otherwise of all forms of euthanasia. He would, for instance, in principle be in favour of *voluntary* euthanasia, provided it applied within narrowly defined limits. Obviously, we do not want to find ourselves putting down hypochondriacs seeking relief from a toothache, but a dying patient whose ebbing life is artificially prolonged and sustained through tubes, catheters and electrodes, and whose consciousness is totally overshadowed by physical distress and mental anguish, has no independent personal volition left to carry on living meaningfully.[80]

We can bracket out the case of the dying patient whose life is artificially prolonged since it need not be euthanasia to withdraw the means of artificial prolongation. One problem raised by the above passage is the question of how voluntary euthanasia is to be requested by a dying patient who has 'no independent personal volition left'. There is also an ambiguity in the final sentence surrounding the possession of personal volition, and 'living meaningfully'. Does this mean that to be ineligible for euthanasia a person must both possess personal volition *and* live meaningfully (which is prevented by being hooked up to machines); or simply that one need only possess personal volition by itself without necessarily being able to use it in a constructive way?

There is something contradictory about volition being used as the test for a meaningful life. If a person has sufficient volitional capacity to make an informed request for euthanasia, it would seem he is in possession of the most 'meaningful aspect' of human existence, namely volition. Since 'any question relating to the quality of life should be measured against the degree of volition that is capable of being exercised',[81] it would appear that any person capable of requesting voluntary euthanasia has a high quality of life and is therefore paradoxically disqualified for it. On this criterion, any request for euthanasia would be self-defeating. There is, however, another ground for euthanasia, namely pain. The following passage sums up van Loon's position more fully regarding the grounds for euthanasia and its permissible forms.

> Therefore, whether a patient is conscious or unconscious, and expressly requests euthanasia or not, and whether he is already in the process of dying or alive but incurably and painfully handicapped, the general rule should be applied that where a disease or disability – or medical treatment itself – induces in a patient either volitionless, unconscious vegetative existence or an overwhelming awareness of distress, pain and suffering to the exclusion of almost any other sensation or conscious activity – then such a patient should be eased into as 'natural' a death as possible, with a minimum of suffering.[82]

It thus appears that van Loon is proposing two grounds for euthanasia: absence of volition, and pain. Only the slenderest justification is offered for euthanasia on grounds of pain. This is found in the statement that 'our immediate moral responsibility is the *relief* of suffering'.[83] Whether or not this is so in Buddhism, and we have advanced reasons in Chapter 1 why it may not be, a good deal of argument would be necessary to show how the alleged injunction to relieve pain could justify the taking of life. Such a conclusion is, of course, directly contrary to the judgement in the case from the Monastic Rule considered in Chapter 1.

The above statement also either contradicts, or advances on, the claim made earlier that Buddhism would be in favour of *voluntary* euthanasia. Above, van Loon suggests that Buddhism endorses both voluntary and non-voluntary euthanasia, since a person may be killed whether he 'expressly requests euthanasia or not'. As regards the different forms of euthanasia, van Loon states that the above definition 'covers what is known as "passive" euthanasia' as well as 'borderline cases of active-passive euthanasia'. The example offered of a borderline case is that of a patient in chronic agony who is eventually killed by the toxicity of the painkilling drugs administered to him. It is not explained whether this case is thought to be 'borderline active' or 'borderline passive'. If the death was an unintended consequence of administering the drugs, of course, it would not really be a case of euthanasia of either kind.

Summary

Van Loon's conclusion seems to be that Buddhism supports active, passive, voluntary and non-voluntary euthanasia. The coma patient should be killed because he has slipped into 'volition death'. The patient who is in chronic pain should be killed (apparently whether he requests it or not) because pain reduces the quality of a person's life to an unacceptable level. Van Loon's contribution is an interesting attempt to make out a case for euthanasia on the basis of the neocortical definition of death. Since Buddhism would not accept this definition of death, however, the case for euthanasia cannot be made on this basis. For Buddhism, the life which is manifested in volition is exactly the same life which is manifested in heartbeat and respiration. There are not two lives, one mental and the other bodily. To lose the capacity for volition is to lose an ability, not to lose one's life.[84]

Other problems have been mentioned above, but there is a final one which should be noted. This is that a definition of death in terms of the higher cognitive faculties is one which can apply only to human beings. Since animals lack the capacity to 'reflect and intuit', some other definition of death will be required in their case. We would thus be in the position of having separate definitions of death for humans and animals, a situation which is especially odd in the context of Buddhism. What is required by Buddhism is a definition of death which will apply to all karmic life. Since different species have different intellectual capacities this must surely be one which makes reference to their underlying organic natures as the basis for determining when death has occurred.

Summary of the three views

The discussion above reveals a number of perspectives on euthanasia. Philip Kapleau seems to regard it as wrong because it is a futile attempt to evade one's karma. While this is certainly so, we would suggest that its wrongness lies not just in its futility. Kapleau offers a different reason in the case of suicide, namely that it deprives one of a human life and 'only with a human body-mind can one become enlightened.'[85] While this is closer to our own approach it is an argument that would have little force in the case, say, of an individual endowed with great karmic merit who immediately gained another human rebirth.

Two reasons were suggested by Phillip Lecso as to why Buddhism regards euthanasia as wrong. The first reason, also mentioned by Philip Kapleau, is that it is an interference with karma. The second is that it is often achieved through increasing doses of narcotics which leave the patient in a comatose condition which deprives him of the mindful understanding of what is to transpire.[86] Neither of these reasons goes to the heart of the Buddhist opposition to euthanasia for reasons given earlier. In brief, the workings of karma are mysterious and we simply cannot know what is part of a person's karma and what is not. It may be that to be killed by euthanasia is itself karmic retribution for an evil done in the past. Clearly, it is impossible to base moral judgements on so speculative a theory. The second argument is not an objection to euthanasia so much as a restatement of the importance Buddhism attaches to a mindful death. As with karma, such an argument could be turned around and used in favour of euthanasia in that a mindful death could be procured before the ravages of a painful illness take hold. Although it is important to die as mindfully as possible, it must be recognised that many people die peacefully, naturally and unconsciously in their sleep, without, one imagines, their spiritual progress being greatly hindered thereby.

The arguments advanced by van Loon in favour of euthanasia depend upon a view of human nature which Buddhism does not accept. These ideas are rejected not only by Buddhism, but also by most ethicists and medical practitioners. Few doctors would be willing to pronounce death on the ground of the higher brain functions being lost while a patient continues to breath spontaneously, and such a patient would certainly not be declared dead according to the brainstem death criteria. As far as Buddhism is concerned we saw that van Loon's arguments have no real basis in terms of traditional Buddhist views of life and death. In fact, his conclusions about Buddhist attitudes to euthanasia are almost diametrically opposed to the views expressed in the sources we have examined.

Conclusion

We note in the literature some confusion about how euthanasia is defined. There is evidence of understandable concern about patients being kept alive as prisoners of technology, and allowing such patients to die rather than prolonging their lives through extraordinary means is sometimes confused with passive euthanasia. It is important to bear in mind here the difference between passive euthanasia and the withholding or withdrawal of burdensome and futile treatment. While Buddhism regards life as a basic good it does not follow that it is something which must be preserved at all costs. Death is a natural part of the *samsaric* cycle and must be accepted as such. Death is not a final end but the doorway to rebirth and new life. The recognition that this is so leads to the abandonment of medical treatment which serves no useful purpose. From the perspective of Buddhist ethics, there is no obligation upon doctors to keep patients alive at all costs. In the case of elderly or terminal cases it is far more important to assist patients in developing the right mental attitude towards death rather than attempting to deny or postpone it. We share the view of Phillip Lecso that a caring environment, such as is offered by the hospice movement, is the type of response which Buddhism would endorse in these circumstances. Where patients are in great pain it may be necessary to administer drugs and other medication although recognising that the quantities involved may shorten the patient's life. The doctor's aim here, however, is to kill the pain, not the patient.

Modern arguments in favour of euthanasia emphasise the principle of autonomy, that is to say, the right of an individual to choose life or death for themselves. What is claimed in practice, however, is the right to have another, usually a doctor exempted by the state from the law of murder, administer euthanasia. The involvement of the doctor, however, changes the issue from one of individual rights to a question of the role of the medical profession in general. The issue is no longer one of individual rights because the doctor is not simply an instrument of the patient's will. The doctor himself must also concur with the patient's reasons for seeking euthanasia before he will administer it. In the normal course of things doctors do not simply carry out the instructions of their patients: they must also use their own professional judgement about what is clinically and ethically right in a given case. A doctor who administers euthanasia has in effect concurred with the patient in the judgement that 'this is a life which is not worth living'. This is a fundamental change in the basis of traditional medical ethics, since physicians are not normally called on to pass judgement as to whether lives are worth living or not before treating illnesses.

In our view Buddhism is opposed to euthanasia essentially because of its affirmative valuation of life. To value death above life by 'making death one's aim', or 'eulogising death', and so forth, is to deny that life is a basic good. The ultimate aim of Buddhism is to overcome death once and for all, and any affirmation of death or choice in favour of death is a rejection of this vision of human good. Since a denial of this kind is central to any form of euthanasia it follows that no form of euthanasia whether active, passive, voluntary, non-voluntary or involuntary can be morally acceptable. A doctor who administers euthanasia is acting as a 'knife-bringer' and thereby doing something which is explicitly prohibited by Buddhist precepts.

We have concentrated on setting out the objections to euthanasia from the Buddhist perspective, and said nothing about other more general objections and grounds for concern. These concerns have been raised elsewhere in the literature on euthanasia and do not need to be repeated in detail here. We might summarise them as the danger of a 'slippery slope' from voluntary euthanasia to non-voluntary euthanasia, the concern that vulnerable people, especially the elderly, would feel themselves under pressure to 'request' euthanasia, and the unknown consequences of changing the role of the medical profession from one which is committed to life to one which connives at death. Where the practice has been officially tolerated and widely practised, there is evidence of abuse.[87] Finally, it is difficult to see how the legalisation of euthanasia, involving the overturning of the traditional Indo-European respect for life, would promote any important public purpose. For these and the other reasons mentioned above Buddhism would oppose the practice of euthanasia in any of its forms, and its legislation.

Notes

INTRODUCTION

1. Beauchamp and Childress (1989:9). Perhaps 'applied *cross-cultural* normative ethics' would be more accurate in the present case. I am not entirely sure which is the genus and which is the species.
2. Buddhists prefer to speak of 'rebirth' rather the 'reincarnation' since the latter implies the existence of an unchanging soul, which is something Buddhism denies.
3. On fundamentalism in religious traditions see Marty and Appleby (1993).
4. King (1964:vf)
5. Answers to the first four questions were proposed in *The Nature of Buddhist Ethics.*

CHAPTER 1: BUDDHISM, MEDICINE AND ETHICS

1. Soni, R.L. (1976:137).
2. Duncan, A.S., G.R. Dunstan, and R.B. Welbourn (1981), *Dictionary of Medical Ethics.* London: Darton, Longman and Todd.
3. Zysk (1991:4). 'This study contends that the traditional account of Indian medicine is merely the result of a later Hinduization process applied to a fundamentally heterodox body of knowledge in order to render it orthodox.' Mitra points out that the Pali canon does not contain the word '*Ayurveda*', but makes reference to all of its traditional branches of treatment (1985:21).
4. Zysk (1991:117).
5. Zysk (1991:6).
6. Demieville (1937:238); MA.i.115.
7 D.i.12.
8. Birnbaum (1979:6).
9. Birnbaum (1979:3f).
10. Trans Zysk (slightly amended) (1991:41).
11. Rock Edict II, trans. Nikam and McKeon (1978:64).
12. Birnbaum (1979:7).
13. Zysk (1991:71).
14. Mizutani (1991:7).
15. A classic early source for this is the 'Discourse on the Fruits of the Religious Life' (*Samaññaphala-sutta*).
16. Dumont (1985:95) (original emphasis).
17. See Eller (1992:102).
18. Wei-hsun Fu (1991:328).

19. Other writers have drawn attention to the positive and negative symbolism of water in Buddhism as a 'stream' in connection with the psychology of the religious life. See Collins (1982:247–61), Gethin (1992:247ff).
20. LaFleur (1992:19), original emphasis.
21. The puritan element in Buddhism should not be overstated, and is counterbalanced by the strong erotic tone and explicit sexual imagery found in the art of many Buddhist cultures. On sex in Buddhism see Stevens (1990).
22. Williams (1989:4).
23. Williams (1989:6).
24. D.iii.124; A.ii.167f; Lamotte (1983–4:10f).
25. Lamotte (1983–4:11).
26. On textual interpretation see Lamotte (1985).
27. VA.i.230. Perhaps a rather forced interpretation of *sutta*, but not of any relevance to the question of the importance of scripture.
28. Vin.i.251.
29. VA.i.231.
30. This follows the chronology suggested by Gombrich (1992).
31. VA.i.231.
32. VA.i.231.
33. VA.i.231.
34. Nett. 22, trans. Lamotte, slightly amended (1983–4:13).
35. Coward (1992).
36. Coward (1992:133).
37. Stout (1990:15).
38. Stout does not mention Buddhism in his book, although there is a passing reference to Hinduism (1990:137).
39. Stout (1990:15).
40. Stout (1990:23).
41. Finnis (1980).
42. Stout (1990:23).
43. Finnis (1980:23).
44. Finnis (1983:70).
45. See *The Nature of Buddhist Ethics*, especially chapter six.
46. Pellegrino (1992:18).
47. Quoted in Lukes (1985:297). Further evidence of the overlap between Indian and Western thought, with particular relevance to Buddhism, can be found in Michael Carrithers' discussion of Mauss's concept of the person. What Carrithers refers to as the *moi* theory bears a close resemblance to natural law. Two of the three moral systems Carrithers mentions as examples of *moi* theories – Stoicism and Christianity – have traditionally been characterised as natural law doctrines. The third example he gives is Buddhism, and we believe Buddhism lends itself readily to such a characterisation.
48. Deepadung (1992:197).
49. Pellegrino (1992:14).
50. Griffiths (1986:xvii).

51. Griffiths (1986:41) sees early Buddhism as holding such a view, and such a reading is by no means implausible.
52. A useful account of the Theravada position may be found in Harvey (1993).
53. A thorough analysis based on early primary sources may be found in Gethin (1986). For an interesting discussion of Buddhist psychology in terms of cognitive science see Varela, Thompson and Rosch (1993).
54. Mahoney (1984:56f).
55. *Essay Concerning Human Understanding*, II.27.9.
56. Mahoney (1984:54).
57. The question of moral personhood in this context raises issues connected with cross-cultural philosophising mentioned earlier. The interest in the criteria of 'personhood' in Western ethics and philosophy has arisen to a large degree because of the importance given to the individual in Western culture, and the question arises as to whether the Western concept or 'category' of the person, with its complex and distinctive social and intellectual history has any Buddhist equivalent. For reasons mentioned above, I think that it has. Several excellent essays exploring this issue may be found in Carrithers, Collins, and Lukes, (eds) (1985).
58. For a discussion on the moral status of 'persons' understood more broadly as 'individuals' in Buddhism see Harvey (1981,1987).
59. Donogan (1977:170).
60. Chapter IX of the *Treasury of Metaphysics* is devoted to a refutation of this thesis.
61. Pm.511.
62. Collins (1982:160).
63. See especially Pillar Edict V and Rock Edict I (Nikam and McKeon, 1978:55f).
64. On this general question see Willson (1987).
65. D.ii.93; M.iii.167. A brief discussion of the Buddhist attitude to animals can be found in Chapple (1992:49–62). See also Story (1964). The belief in rebirth may not be quite so alien to the Western mind as is commonly thought: according to a Gallup poll, 25 per cent of the population of the UK believe in reincarnation (Puttick, 1993:9). On reincarnation and early Christian doctrine see MacGregor (1982).
66. Cf. Harvey (1987:40ff).
67. Vin.iv.125.
68. Vin.i.137.
69. Vin.i.137 trans Wijayaratna (1990:20), slightly amended, emphasis added.
70. D.i.5.
71. D.ii.66f; M.iii.282. The contemporary equivalent of this is found in popular psychotherapies which maintain that the uninhibited expression of emotion is the way to human fulfilment.
72. Roy Perrett (1987) and Padmasiri de Silva (1991:62) assume it does. Obviously, I disagree.
73. On the suggestion that the Buddhist precepts are a form of rule-utilitarianism see *The Nature of Buddhist Ethics*, chapter 5

74. A.i.188f; cf.M.i.415ff.
75. Beauchamp and Childress (1989:6ff).
76. See *The Nature of Buddhist Ethics, passim*.
77. The clearest modern exposition of the principles of natural law ethics is to be found in Finnis (1980). Much of what follows is based on the account given there.
78. On this topic see Collins (1986).
79. Miln 185.
80. For a brief statement of the Buddhist position see Ratanakul (1985:289f).
81. Kleinig (1991:29).
82. Cf. Kleinig (1991:9).
83. Ford (1991:xvf).
84. Harvey (1990:16).
85. Rajavaramuni (1990:36).
86. A.v.87.
87. S.i.88 quoted at Vism 88 (trans. Nyanamoli).
88. S.v.2.
89. D.iii.187. Cf. *Upasakajana-alankara* 267.
90. *Nicomachean Ethics* Penguin Classics (1956 revised edition) p. 261
91. Grisez (1970:304).
92. D.iii.62f.
93. Wiltshire (1983:131).
94. Lamotte (1987).
95. See Sharma (1987).
96. Lamotte (1987:115), ennumeration added.
97. On religious suicide in Christianity and Judaism see Droge and Tabor (1991).
98. Lance Cousins has shown this with respect to the account of the suicide of Channa (Buddha-L Buddhist Academic Discussion Forum, 29 March 1994).
99. Lamotte (1987:114).
100. A good theory should also make accurate predictions. The predictions to be made here would relate to the behaviour of an enlightened subject, but for obvious reasons it is difficult to put the theory to the test in the present day. However, the theory should also predict retrospectively, so to speak, the behaviour of the enlightened as recorded in scripture. This means simply that all examples of moral conduct recorded in mainstream sources should be consistent with the principles set out above.
101. This is one of the four most serious monastic offences for which the penalty is lifelong expulsion from the Order.
102. This is because no case of this kind had arisen hitherto, and the implication of their actions occurred to them *after* the death of the patient.
103. VA.ii.464.
104. David Walker, *The Oxford Companion to Law* (Oxford:Clarendon Press, 1980); 'motive'.
105. Goff (1988:41f), original emphasis.

CHAPTER 2: AT THE BEGINNING OF LIFE

1. This term is used in a non-technical sense to denote the beginning of individual life. It involves no judgement as to *when* life begins.
2. Vism 554 (trans. Nyanamoli). ⌡
3. See, for example Lama Lodo (1987), and Lati Rinbochay and Jeffrey Hopkins (1980).
4. Kelsang (1977:47).
5. Lama Lodo (1987:48).
6. Tsong-kha-pa's *Great Exposition of the Stages of the Path*, quoted in Lati Rinbochay and Jeffrey Hopkins (1980:59).
7. A useful account of Aristotle's views on embryology and their influence may be found in Ford (1991:Ch2).
8. Soni, R. L. (1976:142).
9. Jolly (1951:75).
10. On the related term *'okkamati'* see Collins (1982:208–13).
11. Here *gabbha* or 'embryo' following Buddhaghosa's interpretation.
12. M.i.256.
13. M.ii.156f.
14. Tr. Lipner (1989:54).
15. Kelsang (1977:47).
16. MA.ii.310.
17. Jolly (1951:74). The image of the lotus closing is also used by the Buddhist source, *The Ambrosia Heart Tantra* (Kelsang, 1977:48).
18. M.ii.148.
19. Asl 321; Kelsang (1977:48). Kelsang's Tibetan text mentions a magical means for changing the sex of the embryo (1977:50f). Techniques for accomplishing this were also known to Ayurveda.
20. For example Vism 595.
21. M.ii.17.
22. S.iii.9.
23. D.i.62f.
24. S. i. 206.
25. On these pre-natal stages see S.i.206, Kvu.494, Nd.I.120, SA.i.300.
26. Vism 236. The stages of fetal growth as understood by Tibetan medicine are depicted in the 'Blue Beryl' paintings. See Plate 5 on embryology in the *Middle Way* vol.67,4, February 1993 p. 228ff. On the correspondence with Ayurvedic embryology, Mitra (1985:296ff).
27. Vism 283.
28. Lati Rinbochay and Jeffrey Hopkins (1980). Cf. Donden (1980).
29. Lati Rinbochay and Jeffrey Hopkins (1980:61).
30. Lati Rinbochay and Jeffrey Hopkins (1980:62). Aristotle had a similar idea (Ford, 1991:29).
31. Lati Rinbochay and Jeffrey Hopkins (1980:63).
32. Lati Rinbochay and Jeffrey Hopkins (1980:62).
33. Lati Rinbochay and Jeffrey Hopkins (1980:63).
34. MA ii..310.
35. Kelsang (1977:47).

36. Kelsang (1977:47).
37. Ford (1991:103). Sperm can survive for much longer than this, perhaps for several days.
38. Francis Story, quoted in Kapleau (1972:48), parenthesis added.
39. Ford (1991:105f).
40. Ford (1991:181).
41. Miln 301f.
42. Reading *sosetva* for *sodhetva*.
43. VA.ii.468f.
44. This is the conclusion reached by Ford (1991), who provides much useful scientific data. See especially chapters five and six.
45. Stott (1985:13).
46. Vism 575.
47. DA.ii.509.
48. Ford (1991:133).
49. Iglesias (1990:99).
50. For example M.i.73; D.iii.230; Vism 552.
51. There are examples in the *Jatakas* of pregnancies without intercourse. See Miln 123ff.
52. The moral status of an unanimated embryo is discussed in section five below.
53. Lipner (1989:43).
54. Lipner (1989:49).
55. D.i.11.
56. If the abortion was necessary to save the life of the mother, it seems certain Buddhism would share the view of Hindu jurists that it was morally permissible.
57. JA.iii.121–2.
58. DhA.i.45f.
59. VA.ii. 441.
60. Vin i.97.
61. Vin iii.73.
62. VA.ii.437f.
63. These cases can be found at Vin.iii.83f.
64. VA.ii.455.
65. Harvey (1990:202).
66. Ling (1969:58).
67. Buddhist sources do not discuss either killing in self-defence or judicial execution. However, the former need not involve a breach of the First Precept, and the latter may also be justifiable in Buddhist terms. The purpose of this book is to sketch general principles, and to discuss these 'exceptions' here would involve too much of a digression.
68. MA.i.198.
69. Jones (1989:176).
70. Jones (1989:176f).
71. Jones (1989:177).
72. Jones (1989:176).
73. Stott (1985:8).
74. Lecso (1987:217).

75. 'Abortion' p. 138.
76. Jones (1989:166ff).
77. Hopkins (1988:91).
78. Lama Lodo (1987:41).
79. 'A Shin Buddhist Stance on Abortion' *Buddhist Peace Fellowship Newsletter*, July 1984, p. 5, Buddhist Churches of American Social Issues Committee. I am grateful to Charles Prebish for his assistance in obtaining a copy of this statement.
80. ibid., p. 7.
81. Quoted in Lecso (1987:215).
82. M.iii.118.
83. M.iii.136.
84. See, for example, Buswell (1992:217ff).
85. D.i.12f; cf D.i.22f. In the first case, from the recollection of past lives in meditation the erroneous conclusion is drawn that the world and the soul are eternal. In the second, meditative experience leads to the mistaken conclusion that the world is finite. At D.i.36f the false views are that nirvana is identical with the soul's experience of one or other of the four *jhanas*.
86. Some light is shed on this question by LaFleur (1992).
87. Bardwell Smith (1988:8).
88. Further details can be found in Bardwell Smith (1988) and LaFleur (1992).
89. LaFleur (1992:xiii). Robert Aitken's views on abortion in the light of traditional Japanese beliefs are discussed by LaFleur at pp. 198ff.
90. LaFleur (1992:24).
91. LaFleur (1992:40).
92. LaFleur (1992:11).
93. Miura (1983:26).
94. Miura (1983:26f).
95. Miura (1983:14).
96. Miura (1983:23f).
97. LaFleur (1992:192).
98. Ling (1969:57).
99. Ling (1969:58).
100. Ling (1969:58).
101. Ling (1969:58).
102. Ling (1969:58).
103. Ling (1969:58).
104. LaFleur (1992:116).
105. LaFleur (1992:117).
106. LaFleur (1992:117).
107. Stout (1990).
108. Stout (1990:75).
109. Stout (1990:76).
110. Ling (1969:58).
111. LaFleur (1992:11).
112. At M.i.445 he is asked why there are fewer *Arahats* but more rules, and explains this by reference to the decline of Dhamma. Asoka, too,

comments wistfully in his fifth Rock Edict that immoral conduct has been on the increase 'for many hundreds of years' (Nikam and McKeon, 1959:31).

113. D.iii.7.
114. DA.iii.853
115. This was the theme of Gombrich (1971).
116. Fisher (1989:197f).
117. Fisher (1989:196).
118. Stott (1986:14).
119. Vin iv.261; VA.iv.921.
120. Stott (1985:13).
121. Vin.iii.83. The text actually speaks of 'bearing' children (*vijayati*) rather the 'conceiving' them. However, the contrast drawn in the first case between being barren (*vañjha*) and the aim of the treatment suggests that the issue is about fertility rather than carrying and birthing children.
122. For Buddhaghosa's opinion on which kinds of patients should be treated and how see VA.ii.469ff.
123. Gombrich and Obeyesekere report that attempts to 'Buddhicise' the marriage ceremony are occurring in Sri Lanka, apparently in an attempt to mimic practices in the West. See their *Buddhism Transformed: Religious Change in Sri Lanka* Princeton: Princeton University Press, 1988 p. 267.
124. Sn.26, v.147.
125. Ling (1969:53).
126. Ling (1969:53–60). Ling points out that attitudes to contraception may shift depending on the balance of population between Buddhists and non-Buddhists.
127. Dr Luang Suriyabongs *Buddhism in the Light of Modern Scientific Ideas* (Bangkok, 1960) quoted in Ling (1969:54).
128. Ling (1969:54).
129. Ling (1969:54f).
130. Ling (1969:57).
131. Ling (1969:58).
132. Ling (1969:59).
133. Ling (1969:59).
134. Ling (1969:54).
135. Ling (1969:56).
136. Ling (1969:57).
137. Ling (1969:60).
138. Fitzpatrick (1988:258f)
139. Some 90 per cent of tubal occlusions are caused by previous abortion, the use of the IUD, and sexually transmitted diseases. Pelvic inflammatory disease caused by sexually transmitted infection is a common cause (Iglesias, 1990:146).
140. Megasthenes' opinion is reported by Strabo (Zysk 1991:28).
141. Zysk (1991:31).
142. D.i.12.
143. Fisher (1989:chs 24–6).

144. Fisher (1989:300).
145. Hathout (1992:70).
146. Mahoney (1984:14f).
147. Quoted in Fitzpatrick (1988:264f).
148. See, for example, the *Discourse to Sigala.*
149. Fisher (1989:190).

CHAPTER 3: AT THE END OF LIFE

1. Gervais (1986:2).
2. Gervais (1986:15f).
3. 'A Definition of Irreversible Coma: Report of the Ad Hoc Committee of the Harvard Medical School to Examine the Definition of Brain Death', chairman Henry K.Beecher.
4. Gervais (1986:9).
5. Gervais (1986:8).
6. Gervais (1986:19).
7. Van Loon simply asserts this as the opinion of Buddhism without any consideration of the criteria for a 'Buddhist view'.
8. Van Loon (1978:79).
9. Van Loon (1978:77).
10. Van Loon (1978:78).
11. S.iii.143.
12. M.i.296.
13. M.i.295.
14. Griffiths (1986:12).
15. Griffiths (1986:10).
16. Griffiths (1986:10f).
17. Griffiths (1986:13), emphasis added.
18. M.i.296.
19. MA.ii.351.
20. Cf. Vism 447.
21. VA.ii.439.
22. VA.ii.438.
23. Mettanando (1991:204).
24. *Treasury* IV.73ab.
25. Cf. Miln 306.
26. *Treasury* II.45ab.
27. *Treasury* II.45ab.
28. Mettanando (1991:204).
29. Mettanando (1991:204).
30. Mettanando (1991:205f).
31. Mettanando (1991:206).
32. This shows, incidentally, that van Loon's understanding of *cetana* as 'volition' is defective, although we do not need to pursue this particular aspect of the matter here. *Cetana* is a much more radical psychic function than 'volition' (Keown, 1992:213–22).

33. We say 'apparently' since Mettananda does not make his concept of death explicit, and concentrates instead on the criteria for death. There may, therefore, be no disagreement in this respect.
34. For criticism of the brainstem criteria see Byrne and Nilges (1993).
35. Byrne and Nilges (1993:7).
36. 'Diagnosis of Brain Death. Statement issued by the honorary secretary of the Conference of Medical Royal Colleges and their Faculties in the United Kingdom on 11 October 1976,' *British Medical Journal* 2, 1187–8. This statement was supplemented by a Memorandum in 1979 entitled 'Diagnosis of Death'.
37. 'Memorandum issued by the honorary secretary of the Conference of Medical Royal Colleges and their Faculties in the United Kingdom on 15 January 1979', *British Medical Journal* 1, 332.
38. Becker (1990:543).
39. For a commentary of the legal and ethical issues in this case see Finnis (1993), Keown (1993).
40. [1993] 1 *All England Law Reports* p. 863.
41. Grisez & Boyle (1979:1).
42. [1993] 1 *All England Law Reports* p. 863.
43. Dworkin (1993:192).
44. Mettanando (1991:10).
45. [1993] 1 *All England Law Reports* p. 859.
46. Zysk (1991:73).
47. Early Buddhist sources make reference to a form of treatment for headaches involving the introduction of medicine through the nose (Vin.iii.82).
48. [1993] 1 *All England Law Reports* p. 885.
49. MA.i.198.
50. [1993] 1 *All England Law Reports* p. 865.
51. Vin.iii.70.
52. D.iii.49.
53. Vin.iii.71.
54. Vin.iii.72.
55. Vin.iii.85.
56. Vin.iii.85.
57. Vin. iii.85.
58. A useful sample of Buddhist views on euthanasia may be found in *Raft, The Journal of the Buddhist Hospice Trust* (No.2 Winter, 1989/90). I am grateful to Ray Willis, the Honorary Secretary to the Trust, for kindly providing me with a copy of this issue. Elsewhere, Nakasone accuses the Buddha of making 'contradictory pronouncements' about euthanasia and of 'ambivalence' about the taking of life (1989:66). On euthanasia in Hinduism, with brief references to Buddhism, see Young (1989).
59. Kapleau (1972,1989).
60. Kapleau (1989:132–7).
61. Kapleau (1989:131).
62. Kapleau (1989:135).
63. Kapleau (1989:133).

64. Kapleau (1989:134).
65. It goes without saying that I do not include Roshi Kapleau here.
66. This example is cited in Kapleau (1989:133).
67. This example is taken from Mahoney (1984:45f).
68. Examples of both kinds of cases are given in Kapleau (1989:100ff).
69. Lecso (1986:55).
70. Lecso (1986:55).
71. Lecso (1986:55).
72. Lecso (1986:55).
73. This argument is noted by Harvey (1987:100).
74. Lecso (1986:56).
75. Van Loon (1978:72).
76. Job 14.1.
77. Van Loon (1978:75).
78. Van Loon (1978:77).
79. Van Loon (1978:75).
80. Van Loon (1978:78).
81. Van Loon (1978:75).
82. Van Loon (1978:76).
83. Van Loon (1978:76), original emphasis.
84. Cf. Gormally (1994:125).
85. Kapleau (1989:131).
86. Pain is also mentioned by Kapleau (1989:108–23; 139).
87. Keown, J (1992).

Bibliography

Beauchamp, Tom L. and James F. Childress (1989), *Principles of Biomedical Ethics*, Third edn Oxford: Oxford University Press.

Becker, Carl B. (1990) 'Buddhist views of suicide and euthanasia', *Philosophy East and West* 40, pp. 543–56.

Birnbaum, Raoul (1979), *The Healing Buddha*. Boulder,Colo.: Shambhala.

Buswell, Robert E. Jr (1992), *The Zen Monastic Experience. Buddhist Practice in Contemporary Korea*. Princeton: Princeton University Press.

Byrne, Paul A. and Richard G. Nilges (1993) 'The Brain Stem in Brain Death: A Critical Review', *Issues in Law & Medicine 9*, pp. 3–21.

Carrithers, M., S. Collins, and S. Lukes, eds. (1985), *The Category of the Person. Anthropology, Philosophy, History*. Cambridge: Cambridge University Press.

Chapple, Christopher (1992), 'Nonviolence to Animals in Buddhism and Jainism', in *Inner Peace, World Peace*, ed. Kenneth Kraft, SUNY Series in Buddhist Studies, ed. Matthew Kapstein, Albany, NY: SUNY, pp. 49–62.

Collins, Steven (1982), *Selfless Persons: Imagery and Thought in Theravada Buddhism*. Cambridge: Cambridge University Press.

—— (1986) 'Kalyanamitta and Kalyanamittata', *Journal of the Pali Text Society*, pp. 51–72.

Conference of Medical Royal Colleges (1976) 'Diagnosis of Brain Death. Statement issued by the honorary secretary of the Conference of Medical Royal Colleges and their Faculties in the United Kingdom on 11 October 1976', *British Medical Journal 2*, pp. 1187–8.

—— (1979) 'Diagnosis of Death. Memorandum issued by the honorary secretary of the Conference of Medical Royal Colleges and their Faculties in the United Kingdom on 15 January 1979', *British Medical Journal 1*, p. 332.

Coward, Harold (1992) 'The Role of Scripture in the Self-Definition of Hinduism and Buddhism in India', *Studies in Religion* 21, pp. 129–44.

Deepadung, Attajinda (1992), 'The Interaction between Thai Traditional and Western Medicine in Thailand', in *Transcultural Dimensions in Medical Ethics*, eds Edmund Pellegrino, Patricia Mazzarella, and Pietro Corsi, Frederick, Md: University Publishing Group, pp. 197–212.

Demieville, P. (1937) 'Byo', *Hobogirin* fascicule III.

Donden, Y. (1980), 'Embryology in Tibetan Medicine', in *Tibetan Medicine*, Dharamsala: Library of Tibetan Works and Archives.

Donogan, Alan (1977), *The Theory of Morality*. Chicago: Chicago University Press.

Droge, A.J. and J.D. Tabor (1991), *A Noble Death: Suicide and Martyrdom among Christians and Jews in Antiquity*. San Francisco: HarperCollins.

Dumont, L. (1985), 'A modified view of our origins: the Christian beginnings of modern individualism', in *The Category of the Person. Anthropology, Philosophy, History*, eds M. Carrithers, S. Collins, and S. Lukes, Cambridge: Cambridge University Press, pp. 93–122.

Duncan, A.S., G.R. Dunstan, and R.B. Welbourn (1981), *Dictionary of Medical Ethics*. London: Darton, Longman and Todd.

Dworkin, Ronald (1993), *Life's Dominion*. London: HarperCollins.

Eller, Cynthia (1992), 'The Impact of Christianity on Buddhist Nonviolence in the West', in *Inner Peace, World Peace*, ed. Kenneth Kraft, SUNY Series in Buddhist Studies, ed. Matthew Kapstein, Albany, NY: State University of New York Press, pp. 91–109.

Finnis, J.M. (1980), *Natural Law and Natural Rights*. Clarendon Law Series, H.L.A. Hart, ed. Oxford: Oxford University Press.

—— (1983), *Fundamentals of Ethics*. Oxford: Oxford University Press.

—— (1993) 'Bland: Crossing the Rubicon', *Law Quarterly Review* 109, pp. 329–37.

Fisher, Anthony (1989), *IVF: The Critical Issues*. Blackburn, Victoria: Collins Dove.

Fitzpatrick, F.J. (1988), *Ethics in Nursing Practice. Basic Principles and their Application*. London: The Linacre Centre.

Ford, Norman M. (1991), *When did I begin?* Cambridge: Cambridge University Press.

Gervais, Karen Grandstrand (1986), *Redefining Death*. New Haven: Yale University Press.

Gethin, R.M.L. (1986) 'The Five khandhas: their treatment in the Nikayas and Early Abhidhamma', *Journal of Indian Philosophy* 14, 35–53.

—— (1992), *The Buddhist Path to Awakening: A Study of the Bodhi-Pakkhiya Dhamma*. Brill's Indological Library, Johannes Bronkhorst, ed., vol. 7. Leiden: E.J. Brill.

Goff, Lord Robert (1988) 'The Mental Element in the crime of murder', *Law Quarterly Review* 104, pp. 41–2.

Gombrich, R.F. (1971), *Precept and Practice. Traditional Buddhism in the Rural Highlands of Ceylon*. Oxford: Clarendon Press.

—— (1992), 'Dating the Buddha: A Red Herring Revealed', in *The Dating of the Historical Buddha*, ed. Heinz Bechert, Göttingen: Vandenhoeck & Ruprecht, pp. 237–59.

Gormally, Luke, ed. (1994), *Euthanasia, Clinical Practice and the Law*. London: The Linacre Centre for Health Care Ethics.

Griffiths, Paul J. (1986), *On Being Mindless: Buddhist Meditation and the Mind–Body Problem*. La Salle: Open Court.

Grisez, Germain (1970), *Abortion. The Myths, The Realities, and the Arguments*. New York: Corpus Books.

Grisez, Germain and Joseph M. Boyle (1979), *Life and Death with Liberty and Justice*. Notre Dame and London: University of Notre Dame Press.

Harvey, Peter (1981) 'The Concept of the Person in Pali Buddhist Literature', unpublished PhD Dissertation, University of Lancaster.

—— (1987) 'A Note and Response to "The Buddhist Perspective on Respect for Persons"', *Buddhist Studies Review* 4, pp. 97–103.

—— (1990), *An Introduction to Buddhism*. Cambridge: Cambridge University Press.

—— (1993) 'The Mind–Body Relationship in Pali Buddhism', *Asian Philosophy* 3, pp. 29–41.

Hathout, Hassan (1992), 'Islamic Basis for Biomedical Ethics', in *Transcultural Dimensions in Medical Ethics*, eds Edmund Pellegrino, Patricia Mazzarella and Pietro Corsi, Frederick, Maryland: University Publishing Group, pp. 57–72.

Hopkins, Jeffrey, ed. (1988), *The Dalai Lama at Harvard*. Ithaca, NY: Snow Lion Publications.

Iglesias, Teresa (1990), *IVF and Justice*. London: The Linacre Centre for Health Care Ethics.

Jolly, Julius (1951), *Indian Medicine*. Poona: C.G. Kashikar.

Jones, K. (1989), *The Social Face of Buddhism*. London: Wisdom.

Kapleau, Philip (1972), *The Wheel of Death*. London: George, Allen and Unwin.

—— (1989), *The Wheel of Life and Death*. New York: Doubleday.

Kelsang, Jhampa trans. (1977), *The Ambrosia Heart Tantra. The secret oral teaching on the eight branches of the science of healing. With annotations by Dr. Yeshi Dönden*, Vol. 1. Dharamsala: The Library of Tibetan Works and Archives.

Keown, Damien (1992), *The Nature of Buddhist Ethics*. London: Macmillan.

Keown, J. (1992) 'Euthanasia, Clinical Practice and the Law', *Law Quarterly Review* 108, pp. 51–78.

—— (1993) 'Doctors and Patients: Hard Case, Bad Law, "New" Ethics', *Cambridge Law Journal* 52, pp. 209–12.

King, W.L. (1964), *In the Hope of Nibbana*. La Salle: Open Court.

Kleinig, John (1991), *Valuing Life. Studies in Moral, Political and Legal Philosophy*, Marshall Cohen, ed. Princeton: Princeton University Press.

Kraft, Kenneth, ed. (1992), *Inner Peace, World Peace. Essays on Buddhism and Nonviolence*. SUNY Series in Buddhist Studies, Matthew Kapstein, ed. Albany, NY: State University of New York Press.

LaFleur, William A. (1992), *Liquid Life: Abortion and Buddhism in Japan*. Princeton: Princeton University Press.

Lamotte, E. (1983–4) 'The Assessment of Textual Authenticity in Buddhism', *Buddhist Studies Review* 1, pp. 4–15.

—— (1985) 'The Assessment of Textual Interpretation in Buddhism', *Buddhist Studies Review* 2, pp. 4–24.

—— (1987) 'Religious Suicide in Early Buddhism', *Buddhist Studies Review* 4, pp. 105–26.

Lati, Rinboche and Jeffrey Hopkins (1980), *Death, Intermediate State and Rebirth in Tibetan Buddhism*. Ithaca, NY: Snow Lion Publications.

Lecso, Phillip A. (1986) 'Euthanasia: A Buddhist Perspective', *Journal of Religion and Health* 25, pp. 51–7.

—— (1987) 'A Buddhist View of Abortion', *Journal of Religion and Health* 26, pp. 214–18.

Ling, Trevor (1969) 'Buddhist Factors in Population Growth and Control', *Population Studies* 23, pp. 53–60.

Lipner, Julius J. (1989), 'The Classical Hindu View on Abortion and the Moral Status of the Unborn', in *Hindu Ethics*, eds Harold G. Coward, Julius J. Lipner and Katherine K. Young, Albany, NY: State University of New York Press, pp. 41–69.

Little, David, John Kelsay, and Abdulaziz Sachedina, eds. (1988), *Human Rights and the Conflicts of Culture*. Columbia, SC: University of South Carolina Press.

Lodo, Lama (1987), *Bardo Teachings*. Ithaca, NY: Snow Lion Publications.

Lukes, S. (1985), 'Conclusion', in *The Category of the Person. Anthropology, Philosophy, History*, eds M. Carrithers, S. Collins, and S. Lukes, Cambridge: Cambridge University Press.

MacGregor, Geddes (1982), *Reincarnation as a Christian Hope*. London: Macmillan.

Mahoney, John (1984), *Bioethics and Belief*. London: Sheed and Ward.

Marty, Martin E. and R. Scott Appleby, eds (1993), *Fundamentalisms and Society. The Fundamentalisms Project*. Chicago: University of Chicago Press.

Mettanando, Bhikkhu (1991), 'Buddhist Ethics in the Practice of Medicine', in *Buddhist Ethics and Modern Society: An International Symposium*, eds Charles Wei-hsun Fu and Sandra A. Wawrytko, New York: Greenwood Press, pp. 195–213.

Mitra, Jyotir (1985), *A Critical Appraisal of Ayurvedic Materials in Buddhist Literature (with special reference to Tripitaka)*. Varanasi: The Jyotirlok Prakashan.

Miura, Domyo (1983), *The Forgotten Child*. Henley-on-Thames, England: Aidan Ellis.

Mizutani, Kosho (1991), 'Prologue', in *Buddhist Ethics and Modern Society. An International Symposium*, eds. Charles Wei-hsun Fu and Sandra A. Wawrytko, New York: Greenwood Press, pp. 6–10.

Nakasone, Ronald Y. (1990), *Ethics of Enlightenment*. Fremont, Calif.: Dharma Cloud Publishers.

Nikam, N.A. and Richard McKeon (1978), *The Edicts of Asoka*, Midway Reprint edn Chicago & London: University of Chicago Press.

Pellegrino, Edmund, Patricia Mazzarella, and Pietro Corsi, eds (1992), *Transcultural Dimensions in Medical Ethics*. Frederick, Md: University Publishing Group.

Perrett, Roy (1987) 'Egoism, Altruism and Intentionalism in Buddhist Ethics', *Journal of Indian Philosophy* 15, pp. 71–85.

Puttick, Elizabeth (1993) 'Why has Boddhidharma (*sic*) left for the West? The Growth and Appeal of Buddhism in England', *Religion Today* 8, pp. 5–9.

Rajavaramuni, Phra (1990), 'Foundations of Buddhist Social Ethics', in *Ethics, Wealth, and Salvation*, eds Russell F. Sizemore and Donald K. Swearer, Columbia: University of South Carolina Press, pp. 29–53.

Ratanakul, Pinit (1985), 'The Buddhist Concept of Life, Suffering and Death and their Meaning for Health Policy', in *Health Policy, Ethics and Human Values*, eds. Z. Bankowski and J.H. Bryant, Geneva: CIOMS, pp. 286–95.

Sharma, Arvind (1987) 'Emile Durkheim on Suicide in Buddhism', *Buddhist Studies Review* 4, pp. 119–26.

de Silva, Padmasiri (1991), 'Buddhist Ethics', in *A Companion to Ethics*, ed. Peter Singer, *Blackwell's Companions to Philosophy*, Oxford: Blackwell, pp. 58–68.

Smith, Bardwell (1988) 'Buddhism and Abortion in Contemporary Japan: Mizuko Kuyo and the Confrontation with Death', *Japan Journal of Religious Studies* 15, pp. 3–24.

Soni, R.L. (1976), 'Buddhism in Relation to the Profession of Medicine', in *Religion and Medicine* 3, Vol. 3, ed. D.W. Millard, London: SCM Press, pp. 135–51.

Stevens, John (1990), *Lust for Enlightenment*. Boston and London: Shambhala.

Story, Francis (1964), *The Place of Animals in Buddhism*. Kandy: Buddhist Publication Society.

Stott, David (1985), *A Circle of Protection for the Unborn*. Bristol: Ganesha Press.

Stout, Jeffrey (1990), *Ethics After Babel*. Cambridge: James Clark.

van Loon, Louis H. (1978), 'A Buddhist Viewpoint', in *Euthanasia*, eds Oosthuizen, G.C., H.A. Shapiro, and S.A. Strauss, Human Sciences Research Council Publication No.65, Cape Town: Oxford University Press, pp. 56–79.

Varela, F. J., E. Thompson and E. Rosch (1993), *The Embodied Mind. Cognitive Science and Human Experience*. Cambridge and London: MIT Press.

Wei-hsun Fu, Charles (1991), 'From Paramartha-satya to Samvrti-satya', in *Buddhist Ethics and Modern Society. An International Sympsosium*, eds Charles Wei-hsun Fu and Sandra A. Wawrytko, *Contributions to the Study of Religion*, ed. Henry Warner Bowden, 31, New York: Greenwood Press, pp. 313–29.

Wei-hsun Fu, Charles and Sandra A. Wawrytko, eds (1991), *Buddhist Ethics and Modern Society: An International Symposium*. New York: Greenwood Press.

Wijayaratana, Mohan (1990), *Buddhist Monastic Life*. Cambridge: Cambridge University Press.

Williams, Paul (1989), *Mahayana Buddhism. The Doctrinal Foundations*. London: Routledge.

Willson, Martin (1987), *Rebirth and the Western Buddhist*. London: Wisdom.

Wiltshire, Martin G. (1983) 'The "Suicide" Problem in the Pali Canon', *Journal of the International Association of Buddhist Studies* 6, pp. 124–40.

Young, Katherine K. (1989), 'Euthanasia: Traditional Hindu Views and the Contemporary Debate', in *Hindu Ethics. Purity, Abortion, and Euthanasia*, eds Harold G. Coward, Julius J. Lipner, and Katherine K. Young, McGill Studies in the History of Religions, ed. Katherine K. Young, Albany, NY: State University of New York Press, pp. 71–130.

Zysk, Kenneth G. (1991), *Asceticism and Healing in Ancient India: Medicine in the Buddhist Monastery*. Oxford: Oxford University Press.

Glossary

Abhidhamma	The Scholastic Treatises; the division of the Pali canon dealing with scholastic philosophy
Arahat	One who gains enlightenment by following the teachings of a Buddha
Asoka	Indian Buddhist emperor of the third century BC
Avalokitesvara	A great bodhisattva of Mahayana Buddhism
Ayurveda	Traditional Indian medicine ('science of life')
Bodhisattva	One who seeks enlightenment in order to assist others
Brahmanism	The early classical phase of Hinduism
Brahmin	A member of the highest of the four castes
Buddha	One who gains enlightenment by discovering the truth himself
Buddhaghosa	A famous scholar and commentator of the Theravada tradition (fifth century AD)
Dependent Origination	The doctrine that all phenomena originate from pre-existing conditions. Since this is so, there cannot be an unchanging soul or 'self'
Dhamma	Pali for Dharma (q.v.)
Dharma	Natural law; Buddhist doctrine
Discourses	The section of the Pali canon containing the Buddha's sermons
Dukkha	Suffering, unsatisfactoriness
First Council	A council held shortly after the Buddha's death at which, according to tradition (almost certainly erroneously) the canon was established
Five Precepts	The prohibitions on taking life, stealing, sexual misconduct, lying, and taking intoxicants
Hinayana	A derogatory term implying inferiority in doctrine and practice, applied by the Mahayana to schools such as the Theravada
Kalala	Embryo, zygote
Karma	Moral actions from which good or bad consequences will inevitably flow
Karuna	Compassion
Mahayana	A broadly-based movement which appears around the time of Christ, characterised by more liberal attitudes to doctrine and some aspects of practice. Traditionally prominent in North Asia
Mizuko	Japanese notion of the fetus as 'water child'
Mizuko Kuyo	Religious service held following abortions in Japan
Monastic Rule	The section of the Pali canon dealing with monastic discipline
Nirvana	The *summum bonum*; the end of cyclic existence

Pali canon	The canon of the Theravada school, written in the Pali language. It contains three main divisions: the Discourses, the Monastic Rule, and the Scholastic Treatises
Pañña	Intuitive wisdom, *sophia*
Parajika	'Defeat'. The most serious category of offence in the monastic Rule, punished by excommunication
Prajña	The Sanskrit equivalent of *Pañña*
Prana	Breath, life; the bodily 'humour' regulating the vital functions of life
Pratyekabuddha	A Buddha who does not teach
Samana	A religious mendicant
Samsara	Cyclic existence
Sangha	The Buddhist monastic Order
Sutta	A religious discourse
Theravada	The oldest surviving school of Buddhism, prominent in South-East Asia
Vedic	Pertaining to the Veda, the ancient scriptures of Brahmanism dating to around 1200 BC
Vinaya	The Monastic Rule
Viññana	Consciousness. The fifth category of human nature

Subject Index

Name Index